DRONES AND THEIR APPLICATIONS

Matthew N. O. Sadiku, Ph.D., P.E.

Regents Professor Emeritus and IEEE Life Fellow
Prairie View A&M University
Prairie View, TX 77446
Email: sadiku@ieee.org
Web: www.matthew-sadiku.com

Copyright © 2025 by Matthew N. O. Sadiku, Ph.D., P.E. 814393

All rights reserved. No part of this book may be reproduced or transmitted in any form or by any means, electronic or mechanical, including photocopying, recording, or by any information storage and retrieval system, without permission in writing from the copyright owner.

To order additional copies of this book, contact:
Xlibris
844-714-8691
www.Xlibris.com
Orders@Xlibris.com

ISBN:	Softcover	979-8-3694-3758-2
	Hardcover	979-8-3694-3759-9
	EBook	979-8-3694-3757-5

Print information available on the last page

Rev. date: 04/14/2025

DEDICATED TO MY BROTHER:
JULIUS O. SADIKU

BRIEF TABLE OF CONTENTS

Chapter 1   Introduction 1

Chapter 2   Drones in Agriculture 14

Chapter 3   Drones in Business 27

Chapter 4   Healthcare Drones 43

Chapter 5   Drones in Education 55

Chapter 6   Drones in Manufacturing 66

Chapter 7   Drones In Construction 76

Chapter 8   Drones in Oil & Gas Industry 85

Chapter 9   Drones in Power Systems 98

Chapter 10 Drones in Telecommuncations 109

Chapter 11 Drones in Law Enforcement 121

Chapter 12 Drones in Entertainment 135

Chapter 13 Drones in Surveillance 147

Chapter 14 Drones in Space Exploration 159

Chapter 15 Drones in the Military 171

PREFACE

Technological advances are changing the world around us. Drones are the next wave of technological advance that can make a huge impact on almost all industries. Drones have become widely used for both recreational and commercial purposes worldwide. Specialists in many industries are exploiting the unique flexibility and observational capabilities of drones to improve industrial processes and operational efficiency. Drones have evolved in the past few years to become exceptionally versatile across practically every industry. The reason drones are being adopted by various industries is because they make good business sense and for their multitasking capabilities.

A drone, also known as unmanned aerial vehicle (UAV), is a pilotless aircraft. It may be regarded as a small aircraft that can fly without a human pilot, usually made of lightweight materials, that can be remotely controlled or fly autonomously. Drones, previously used for military purposes, have started to be used for civilian purposes since the 2000s. They have come in a great diversity of several civilian applications. These industries are commonly using drones to improve operations, increase efficiency, and save time and cost on data collection. Drones have proven themselves unique and powerful assets through all their work. As a result, all industries are using them today to get various tasks done.

This book explores the various applications of drones. The book is organized into fifteen chapters that summarize the applications such as construction, agriculture, healthcare, power systems, education, business, manufacturing, oil & gas, telecommunications, entertainment, law enforcement, space exploration, surveillance, and the military.

Chapter 1: Introduction: This chapter explores the various applications of drones and serves as an introduction to the entire book. Drones, designed with a kite-like mechanical architecture, mainly consist of four basic components: propeller, engine, body, and flight board. They make it possible to collect data and access information in a variety of ways while saving time and minimizing costs. As the technology and equipment become more accessible, drones will be more active part of commercial and industrial operations.

Chapter 2: Agriculture: This chapter provides an introduction on the use of drones in agriculture. Agriculture represents the primary food source of the world. The agriculture industry has embraced drones as indispensable tools for farmers around the world. They are helping farmers to optimize agriculture operations, increase productivity, increase crop production, monitor crop growth, reduce pollution, reduce wastage, and reduce time spent in the field. The number of farmers adopting drones in their farms is growing.

Chapter 3: Business: This chapter introduces readers the various applications of drones in the business world. Technology has had a huge influence on workplace productivity. Drones are no exception. Drones have quickly become an effective tool for businesses of all sizes. They have gained popularity in recent years for both business and personal use. Many business executives are quick to realize the myriad of benefits drones could offer by integrating technology into their operations. They are discovering ways to use drones to gather intelligence on the competition. Drones are saving companies time and money.

Chapter 4: Healthcare: In this chapter, we discuss the use of drones in the field of healthcare. Drones are increasingly being used as innovative tools for healthcare. They are emerging as a new healthcare tool that can help mitigate logistical problems and make healthcare more accessible and save lives. Drones could ultimately be a healthcare technology game-changer. The days are fast approaching when cars and trucks will be replaced by drones for moving things across hospital campuses.

Chapter 5: Education: This chapter explores the various applications of drones in education. Drones are naturally fun and educational. Drones, once known as outdoor toys for kids and teenagers, turned out to have immense potential to transform. It has become increasingly common in recent years for teachers to utilize drones as learning tool within the classroom. Incorporating drone information and exercises into the curriculum for STEM students is important for career preparation because of the widespread integration of drones as tools across many fields.

Chapter 6: Manufacturing: This chapter delves into the applications of drones in the manufacturing sector. They have quickly become indispensable tools in the manufacturing industry and are revolutionizing the way manufacturing facilities operate. Their integration into the manufacturing industry offers immense potential for improving efficiency, safety, and cost savings. Drones will continue to play a significant role in shaping the future of manufacturing.

Chapter 7: Construction: This chapter explores the applications of drones in the construction industry. Drones offer immense possibilities and applications in the world of construction, offering benefits ranging from on-site safety to remote monitoring. Their current capabilities allow them to cut costs, time, risk, and labor, while improving workflow, accuracy, communication, and efficiency. While the impact of drones in the construction industry is evolving, they are already revolutionizing the sector and changing the way construction projects are handled.

Chapter 8: Oil & Gas Industry: In this chapter, we examine the applications of drones in the oil and gas industry. The oil and gas industry, known for its vast operations and extensive infrastructure, is undergoing a technological renaissance with the integration of drone inspections. Drones have become a crucial part of the oil and gas industry, where they are being deployed to perform a wide variety of tasks. Drones in oil and gas have helped reduce inspection time, cut costs, decrease downtime, and identify problems early on. When it comes to oil and gas industry, drones are a super significant investment.

Chapter 9: Power Systems: This chapter considers the applications of drones in power systems. The power sector is one of the world's most rigid industries. As a result, the use of drones in the power industry is relatively new, with many companies still unsure about the value the drones could bring to their operations. Utilities are increasingly using drones as a safe and effective tool to assist them in their operations. Drones drastically cut the costs of power line inspections for utilities. They also improve safety, increase reliability, and reduce response time across transmission and distribution systems.

Chapter 10: Telecommunications: This chapter covers the applications of drones in the telecommunications industry. Telecommunications industry reaches into every corner of our economies, societies, and private lives, and it is one of the greatest drivers of economic growth and human equality. It encompasses a wide array of services and technologies, including telephone services, Internet access, and broadcasting. Drone technology is changing the face of telecommunications quickly and its first real application in telcos is tower inspections. As drones in telecommunications industry become increasingly prevalent, they are transforming the way companies approach tower inspections, network expansions, and emergency response.

Chapter 11: Law Enforcement: This chapter highlights the applications of drones in law enforcement. The use of drones by law enforcement agencies has grown exponentially in the past few years. Drones are rapidly being adopted by police departments and law enforcement agencies all over the world, making their work significantly efficient, safer, and easier. In any major city you will find drones supporting the work of police departments, fire departments, search and rescue teams, and disaster management personnel. Drones provide law enforcement with a bird's-eye view of unfolding situations in real time, allowing officers to gather critical information without putting themselves in harm's way.

Chapter 12: Entertainment: This chapter discusses the applications of drones in entertainment. Since the establishment of the film industry, storytellers, screenwriters, and producers have always looked for ways to make movies all the more enjoyable. In recent years, drones have emerged as powerful tools in entertainment industry, revolutionizing the way movies, TV shows, music videos, and live events are shot. Drones are now changing the way movie makers operate. They are also carving out a space for themselves in the film industry that is entirely separate from traditional Hollywood fare.

Chapter 13: Surveillance: This paper highlights the applications of drones in surveillance. Surveillance is the close observation of a person, group of people, activities, infrastructure, building, etc. for the purpose of managing, influencing, directing, or protecting. There are several different methods of surveillance. Drone surveillance is the use of unmanned aerial vehicles (UAVs) to capture still images and video at a high altitude to gather information about specific targets. It presents an easier, faster, and cheaper method of data collection. It can survey objects that may be out of reach and can get a first-person view that photographers do not usually get.

Chapter 14: Space Exploration: This chapter explores major applications of drones in space exploration. The vast expanse of outer space has long captured the imagination of humanity. In recent years, there has been a tendency to design and develop concepts of drones for space exploration. Drones have the potential to revolutionize future space missions, enabling greater understanding and exploration of the universe. Using drones to explore the other planets or moons is one of the main priorities of space agencies. As technology continues to advance, we can only expect the role of drones in space exploration to expand further.

Chapter 15: Military: In this chapter, we cover the applications of drones in the military. Drones have rapidly evolved into an essential component of modern warfare. Today, drones primarily serve the military industry around the world. Military drones have become indispensable tools for special operations, providing real-time situational awareness, communication relay, and electronic warfare capabilities. Their applications have revolutionized the way modern warfare is conducted. From surveillance to combat and logistics, drones have become an indispensable tool for military operations worldwide. Every military specialist agrees that drones are the future of warfare.

This is a comprehensive introductory text on the issues, ideas, theories, and applications of modern drones. It provides an overview of each application so that beginners can understand it. It is a must-read book for anyone who wants to learn about modern drones, which have become vital to all industries in recent times.

I am grateful for the support of Dr. Annamalia Annamalai, the department head of the Department of Electrical and Computer Engineering, and Dr. Pamela Obiomon, the dean of the College of Engineering at Prairie View A&M University, Prairie View, Texas. Special thanks are due to my wife Dr. Janet Sadiku for helping in various ways.

- M. N. O. Sadiku

ABOUT THE AUTHOR

Matthew N. O. Sadiku received his B. Sc. degree in 1978 from Ahmadu Bello University, Zaria, Nigeria and his M.Sc. and Ph.D. degrees from Tennessee Technological University, Cookeville, TN in 1982 and 1984 respectively. From 1984 to 1988, he was an assistant professor at Florida Atlantic University, Boca Raton, FL, where he did graduate work in computer science. In total, he received seven college degrees. From 1988 to 2000, he was at Temple University, Philadelphia, PA, where he became a full professor. From 2000 to 2002, he was with Lucent/Avaya, Holmdel, NJ as a system engineer and with Boeing Satellite Systems, Los Angeles, CA as a senior scientist. He is presently a Regents professor emeritus of electrical and computer engineering at Prairie View A&M University, Prairie View, TX.

He is the author of over 1,330 professional papers and over 140 books including "Elements of Electromagnetics" (Oxford University Press, 7th ed., 2018), "Fundamentals of Electric Circuits" (McGraw-Hill, 7th ed., 2020, with C. Alexander), "Computational Electromagnetics with MATLAB" (CRC Press, 4th ed., 2019), "Principles of Modern Communication Systems" (Cambridge University Press, 2017, with S. O. Agbo), and "Emerging Internet-based Technologies" (CRC Press, 2019). In addition to the engineering books, he has written Christian books including "Secrets of Successful Marriages," "How to Discover God's Will for Your Life," and commentaries on all the books of the New Testament Bible. Some of his books have been translated into French, Korean, Chinese (and Chinese Long Form in Taiwan), Italian, Portuguese, Spanish, German, Dutch, Polish, and Russian.

He was the recipient of the 2000 McGraw-Hill/Jacob Millman Award for outstanding contributions in the field of electrical engineering. He was also the recipient of Regents Professor award for 2012-2013 by the Texas A&M University System. He is a registered professional engineer and a life fellow of the Institute of Electrical and Electronics Engineers (IEEE) "for contributions to computational electromagnetics and engineering education." He was the IEEE Region 2 Student Activities Committee Chairman. He was an associate editor for IEEE Transactions on Education. He is also a member of Association for Computing Machinery (ACM). His current research interests are in the areas of computational electromagnetic, computer science/networks, engineering education, and marriage counseling. His works can be found in his autobiography, "My Life and Work" (Author's Tranquility Press, 2024) or his website: www.matthew-sadiku.com. He currently resides with his wife Janet in Westlake Florida. He can be reached via email at sadiku@ieee.org

DETAILED TABLE OF CONTENTS

Preface vii
About the Author xi

CHAPTER 1	INTRODUCTION 1
	1.1 INTRODUCTION 1
	1.2 WHAT IS A DRONE? 2
	1.3 TYPES OF DRONES 3
	1.4 APPLICATIONS 6
	1.5 BENEFITS 11
	1.7 CHALLENGES 11
	1.8 CONCLUSION 12
	REFERENCES 12
CHAPTER 2	DRONES IN AGRICULTURE 14
	2.1 INTRODUCTION 14
	2.2 WHAT ARE DRONES? 15
	2.3 WHY USE AGRICULTURE DRONES? 16
	2.4 TYPES OF DRONES 17
	2.5 APPLICATIONS 21
	2.6 BENEFITS 22
	2.7 CHALLENGES 24
	2.8 CONCLUSION 25
	REFERENCES 25
CHAPTER 3	DRONES IN BUSINESS 27
	3.1 INTRODUCTION 27
	3.2 WHAT ARE DRONES? 28
	3.3 APPLICATIONS 29
	3.4 BENEFITS 33
	3.5 CHALLENGES 34
	3.7 GLOBAL USES OF DRONES IN BUSINESS 35
	3.8 CONCLUSION 39
	REFERENCES 40

CHAPTER 4	HEALTHCARE DRONES	43
	4.1 INTRODUCTION	43
	4.2 CONCEPT OF DRONES	44
	4.3 APLICATIONS	46
	4.4 GLOBAL HEALTHCARE DRONES	49
	4.5 BENEFITS	50
	4.6 CHALLENGES	51
	4.7 CONCLUSION	52
	REFERENCES	53
CHAPTER 5	DRONES IN EDUCATION	55
	5.1 INTRODUCTION	55
	5.2 WHAT IS A DRONE?	56
	5.3 APPLICATIONS OF DRONES IN EDUCATION	57
	5.4 APPLICATIONS IN SCHOOLS	59
	5.5 BENEFITS	61
	5.6 CHALLENGES	63
	5.7 CONCLUSION	63
	REFERENCES	64
CHAPTER 6	DRONES IN MANUFACTURING	66
	6.1 INTRODUCTION	66
	6.2 WHAT IS A DRONE?	67
	6.4 BENEFITS	71
	6.5 CHALLENGES	72
	6.6 CONCLUSION	74
	REFERENCES	74
CHAPTER 7	DRONES IN CONSTRUCTION	76
	7.1 INTRODUCTION	76
	7.2 WHAT IS A DRONE?	77
	7.3 APPLICATIONS	79
	7.4 BENEFITS	81
	7.5 CHALLENGES	82
	7.6 CONCLUSION	83
	REFERENCES	84
CHAPTER 8	DRONES IN OIL & GAS INDUSTRY	85
	8.1 INTRODUCTION	85
	8.2 WHAT IS A DRONE?	86
	8.3 OIL&GAS DRONES	88
	8.4 APPLICATIONS	89
	8.5 BENEFITS	92

	8.6 CHALLENGES	94
	8.7 CONCLUSION	96
	REFERENCES	96
CHAPTER 9	DRONES IN POWER SYSTEMS	98
	9.1 INTRODUCTION	98
	9.2 WHAT IS A DRONE?	99
	9.3 APPLICATIONS	101
	9.4 BENEFITS	104
	9.5 CHALLENGES	106
	9.6 CONCLUSION	107
	REFERENCES	107
CHAPTER 10	DRONES IN TELECOMMUNCATIONS	109
	10.1 INTRODUCTION	109
	10.2 WHAT IS A DRONE?	110
	10.3 TELECOMM DRONES	112
	10.4 APPLICATIONS	113
	10.5 BENEFITS	115
	10.6 CHALLENGES	118
	10.7 CONCLUSION	119
	REFERENCES	119
CHAPTER 11	DRONES IN LAW ENFORCEMENT	121
	11.1 INTRODUCTION	121
	11.2 WHAT IS A DRONE?	122
	11.3 POLICE DRONES	124
	11.4 APPLICATIONS	126
	11.5 BENEFITS	129
	11.6 CHALLENGES	131
	11.7 CONCLUSION	132
	REFERENCES	133
CHAPTER 12	DRONES IN ENTERTAINMENT	135
	12.1 INTRODUCTION	135
	12.2 WHAT IS A DRONE?	136
	12.3 ENTERTAINMENT DRONES	138
	12.4 APPLICATIONS	138
	12.5 BENEFITS	142
	12.6 CHALLENGES	144
	12.7 CONCLUSION	145
	REFERENCES	145

CHAPTER 13	**DRONES IN SURVEILLANCE**	**147**
	13.1 INTRODUCTION	147
	13.2 WHAT IS A DRONE?	148
	13.3 SURVEILLANCE DRONES	150
	13.4 APPLICATIONS	152
	13.5 BENEFITS	154
	13.6 CHALLENGES	155
	13.7 CONCLUSION	157
	REFERENCES	157
CHAPTER 14	**DRONES IN SPACE EXPLORATION**	**159**
	14.1 INTRODUCTION	159
	14.2 WHAT IS A DRONE?	160
	14.3 SPACE DRONES	162
	14.4 APPLICATIONS	163
	14.5 BENEFITS	166
	14.6 CHALLENGES	167
	14.7 CONCLUSION	169
	REFERENCES	169
CHAPTER 15	**DRONES IN THE MILITARY**	**171**
	15.1 INTRODUCTION	171
	15.2 WHAT IS A DRONE?	172
	15.3 MILITARY DRONES	174
	15.4 APPLICATIONS	177
	15.5 BENEFITS	180
	15.6 CHALLENGES	181
	15.7 CONCLUSION	182
	REFERENCES	183
INDEX		**185**

CHAPTER 1

INTRODUCTION

"Drones, with their agility and small size, seem perfect for search and rescue operations."

- Grant Imahara

1.1 INTRODUCTION

For decades, drones have been used by the military and government organizations to gather data. Their use in the private and commercial sectors is much more recent. Drones have come in a great diversity of several applications such as military, construction, agriculture, healthcare, search and rescue, parcel delivery, hidden area exploration, power line monitoring, wireless communication, and aerial surveillance. These industries are commonly using drones to improve operations, increase efficiency, and save time and cost on data collection.

Drones, also known as unmanned aerial vehicles (UAVs), are unmanned aircrafts. They are small remotely controlled aerial vehicles. They have been crucial to the operations of many enterprises and governmental organizations in recent years, including monitoring the Earth's surface, agriculture, construction, and surveillance. These robot-like aircrafts can assist in the hunt for hurricane survivors as well as deliver groceries to your home and almost everywhere in between. One of their strengths is the many different applications for which they can be used [1].

This chapter explores the various applications of drones and serves as an introduction to the entire book. It begins with describing what a drone is. It mentions different types of drones. It presents some applications of drones. It highlights some benefits and challenges of drones. The last section concludes with comments.

1.2 WHAT IS A DRONE?

The FAA defines drones, also known as unmanned aerial vehicles (UAVs), as any aircraft system without a flight crew onboard. Drones include flying, floating, and other devices, including unmanned aerial vehicles (UAVs), that can fly independently along set routes using an onboard computer or follow commands transmitted remotely by a pilot on the ground. Drones can range in size from large military drones to smaller drones. Drones, previously used for military purposes, have started to be used for civilian purposes since the 2000s. Since then, drones have continued to be used in intelligence, aerial surveillance, search and rescue, reconnaissance, and offensive missions as part of the military Internet of Things (IoT). Today, drones are used for different purposes such as aerial photography, surveillance, agriculture, entertainment, healthcare, transportation, law enforcement, etc.

Drones, designed with a kite-like mechanical architecture, mainly consist of four basic components: propeller, engine, body, and flight board. As shown in Figure 1.1, drones come in two main types: fixed-wing and rotary [2]. In the fixed-wing platforms, the wings are fixed to body of the drone. A rotary platform has a rotary wing (i.e., a propeller) that is fixed to a motor. On a rotary platform, the motor spins the rotary wing against the air to create lift. Also called rotor or blade, drone propeller comes in various shapes, sizes, and materials. The drone motor is the main dynamo that makes the propellers spin and provides enough thrust for flight. Motors and propellers are the technology that lifts the drone, allowing it to fly or hover. The battery is regarded as the heart of the drone; it is the most important part in terms of power generation and performance. The drone controller is the tool for managing and directing the drone. Not all drones have a camera system, but models offer additional options in-flight, such as aerial photography. Drones can operate in Global Navigational Satellite Systems (GNSS) such as GPS [3].

Figure 1.1 Fixed-wing and rotary types of drones [2].

Drones work much like other modes of air transportation, such as helicopters and airplanes. When the engine is turned on, it starts up, and the propellers rotate to enable flight. The motors spin the propellers and the propellers push against the air molecules downward, which pulls the drone upwards. Once the drone is flying, it is able to move forward, back, left, and right by spinning each of the propellers at a different speed. Then, the pilot uses the remote control to direct its flight from the ground [2].

Drone laws exist to ensure a high level of safety in the skies, especially near sensitive areas like airports. They also aim to address privacy concerns that arise when camera drones fly in residential areas. These include the requirement to keep your drone within sight at all times when airborne. In the United States, drones weighing less than 250g are exempt from registration with civil aviation authorities. If your drone exceeds 250g in weight, you will also require a Flyer ID, which requires passing a test [7]. It is necessary to register as an operator, be trained as a pilot, and have civil liability insurance, in addition to complying with various flight regulations, and those of the places where their use is permitted.

1.3 TYPES OF DRONES

Different drones may travel at different heights and distances. Drones can be classified according to size, from very small to large drones. For a drone to fly, it must have a power source such as batteries or fuel. Their power sources classify drones as battery-powered, gasoline-powered, hydrogen fuel cell, and solar drones. They can also be classified as single rotor (helicopter), multi-rotor (multicopter), fixed wing, and fixed-wing hybrid VTOL according to their physical structures. Multicopters are further divided into models with four engines (quadcopter), six engines (hexacopter), and eight engines (octocopter) according to the number of motors. Thus, we have the following types of drones [3]:

- *Tricopter:* It is a type of drone that can take off and land vertically and has six degrees of freedom on the X–Y–Z axes. Although its cost is lower than other options, it has a disadvantage of not being symmetrical (Figure 1.2),

Figure 1.2 Tricopter [3].

- *Quadcopter:* It is the most preferred type of drone. It is simple as well as versatile. It has four propellers and four motors, with a load of up to 5 kg. A higher flight comfort can be achieved with its four-arm structure being symmetrical. In case of any malfunction, the drone will most likely crash (Figure 1.3).

Figure 1.3 Quadcopter [3].

- *Hexacopter:* It is a type of drone with six propellers. It is a type of drone that can offer excellent performance even on indoor flights. It can be equipped with various equipment and take off with a load of up to 10 kg, as in other models (Figure 1.4).

Figure 1.4 Hexacopter [3].

- *Octocopter:* It is a type of drone with eight propellers. It is an advanced type of drone that can take off with a load of 25 kg with its equipment. It is especially preferred for heavily loaded works (Figure 1.5).

Figure 1.5 Octocopter [3].

- *Fixed-wing drones:* Unlike rotary wings, they use wings like a regular airplane instead of vertical lift rotors to provide lift (Figure 1.6). They are much more efficient as they do not use additional power to stay in the air, so they can cover longer distances and scan much larger areas. The main disadvantage of a fixed-wing aircraft is that they cannot fly in one spot. Depending on their size, they need a runway or launcher to get them into the air.

Figure 1.6 Fixed-wing drone [3].

When choosing your ideal drone, budget is the obvious place to start. Then, keep size and weight in mind. Beginner fliers should consider drones with safety features like obstacle avoidance, which help to prevent mid-air collisions.

1.4 APPLICATIONS

Being versatile technology, drones are used across a wide range of industries. The following are ten typical applications of drones [3,4-8]:

1. *Agriculture:* The use of drones in agricultural production is becoming more common daily, and it provides convenience for producers to optimize production. In agriculture, drone contributes significantly to providing data to sector stakeholders and increasing producers' productivity by collecting data with regular land observation. Drones are used in spraying, fertilization, and plant damage detection. Drones can map regions and determine crop damage after a storm, discover water drainage issues, determine quantity of yield, find disease or pest, and check up on livestock quickly. By collecting and analyzing data on soil conditions, irrigation, and plant health, drone technology enables farmers to identify issues early and take corrective action. Figure 1.7 shows how drone is being used in Agriculture [6].

Figure 1.7 Drone is used in Agriculture [6].

2. *Healthcare:* The use of drones for healthcare purposes is used to transport response equipment to the scene in urban areas. Drones are used to improve emergency response after a disaster or in dangerous situations. Medical samples, medications, and supplies can be transported quickly and efficiently by drones, saving time and potentially saving lives. Delivering a life-saving defibrillator (AED) by drone is 32% faster in urban centers and 93% faster in rural areas where other vehicles cannot reach. Drones are also being used to transport donated organs to transplant

recipients. In the same way, it has been demonstrated that it is possible to transport drugs/tissues, and blood products, with drones.

3. *Law Enforcement*: Drones are used for maintaining the law. They help with the surveillance of large crowds and ensure public safety. They assist in monitoring criminal and illegal activities. Law enforcement agencies are finding new and innovative ways to utilize drones in their operations, enhancing public safety and officer efficiency. Drones can also be used for crowd monitoring during large events or protests, enabling law enforcement to identify potential issues and respond accordingly. Drones have been used for domestic police work in Canada and the United States. Many police departments in India have procured drones for law and order and aerial surveillance. Figure 1.8 shows how police officers are using a drone in their operations [7].

Figure 1.8 Police officers are using a drone in their operations [6].

4. *Disaster Management:* Drones have become indispensable tools in search and rescue operations, significantly improving the speed and efficiency with which missing persons can be located. After a natural or man-made disaster, drones can be used to gather information and navigate debris. Its high definition cameras, sensors, and radars give rescue teams access to a higher field of view, saving the need to spend resources on manned helicopters. Due to their small size, drones can provide a close-up view of areas. In situations where time is of the essence, this rapid response capability can mean the difference between life and death. Drones can be used to deliver critical supplies, such as food, water, or medical equipment, to stranded individuals, ensuring their safety and comfort until rescue teams can reach them.

5. *Photography:* Drones provide a detailed view of large areas, spaces, and a particular subject. They have been a benefit to aerial photographers who employ them to get expansive shots. They are most commonly known for capturing stunning aerial photography and videos that are used in marketing or advertising. Professional video shoots are made today using drones in commercials, TV series, and movie sets, successfully capturing specific images. Drones with high-definition cameras serve successfully in aerial image and video shooting at sports events. They provide great convenience in collecting images and information from places that cannot be visited or entered, especially due to security problems. The use of drones can make a significant contribution to innovation and quality in the film industry.
6. *Power Systems:* Drones have proven to be invaluable tools for inspecting and maintaining electric power lines. Power grids sometimes run through areas that are hard to reach and, therefore, drones are the perfect solution to inspect and service these grids. A long range drone equipped cameras and GPS can carry out autonomous cruise along the power grid, and transmit and shoot images in real-time. Drones can safely fly close to transmission lines, towers, and antennas, capturing detailed images and data for analysis. The cost of manual positioning of power line faults is too high, and drone cruises can significantly improve efficiency. Several energy companies, like Southern Company and Duke Energy, are using drones to inspect power lines, power plants, and storm damage. Figure 1.9 shows an example of how drone is monitoring the power system [7].

Figure 1.9 Drone is monitoring the power system [7].

7. *Business:* Drones, modern camera technology, GIS, and specialized software have enabled businesses to leverage new kinds of data for stronger insights. When more precise and consistent measurements and data are provided, businesses have the tools to prepare better financial data. From a business point of view, the drone revolution is leading to an in-depth transformation of the main sectors of activity, since they can take on complex tasks and reduce costs.

8. *Manufacturing:* Integrating drones in manufacturing adds a new standard of automation and efficiency. Manufacturing drones use various advanced technologies that increase their technical capabilities. Drones can be used for the additive manufacturing of structures. Artificial intelligence (AI) plays a critical role in drone use for manufacturing development. AI-driven drones autonomously capture, process, and analyze aerial data. AI is the fundamental technology for creating advanced drones. It makes drones in manufacturing more accessible. One of the most prominent applications of drones in manufacturing is inventory and production line management. There are several tasks that drones can automate in warehouses and production lines. Figure 1.10 shows the use of drone in manufacturing [8].

Figures 1.10 Use of drone in manufacturing [8].

9. *Military:* The first known use of drones was for military purposes. Thus, military is the oldest, most well-known, and most contentious application of drones. Drone is used for different purposes ranging from military, espionage, radar system, area detection, and observation to transporting food, weapons, and ammunition. In the early 1940s, the British and American forces began utilizing extremely crude kinds of drones to spy on the Axis powers. MQ-9 Reaper is one of the drones used for military purposes today. It has a length of 36 feet and is equipped with a 1852 km flight system at an altitude of 50,000 feet. It is outfitted with a variety of missiles and intelligence gathering systems. Islamic State of Iraq and the Levant have used drones to drop explosives, primarily using quadcopters.
10. *Space:* Drones have already conquered the earth's surface and are now heading into space. NASA and the United States Air Force have been testing drones designed for space flight. The Air Force's ultra-secretive X-37B UAV has been quietly circling the Earth for the past two years, setting a record for the longest unmanned aircraft flight (781 days and counting).

Although the Air Force has been ambiguous, it has stated that "the primary objectives of the X-37B are twofold: reusable spacecraft technologies for America's future in space and operating experiments that can be returned to, and examined on Earth."

11. *Entertainment:* Drones are being developed to provide entertainment, as they provide many customizable solutions that combine practicality and speed. Artificial drone intelligence is used in several ways to capture videos and photographs. Thanks to high-definition cameras that record and transmit real-time video of a location while flying at a high altitude. Drones have even made their way into our homes, where they serve as entertainment. As technology advances, drones will become more robust and advanced, accommodating longer flight times and heavier loads

12. *Logistics:* Drones are used in the logistics industry to detect damage and cracks in the ship structure and hull, allowing emergency teams such as the fire brigade to intervene in dangerous areas quickly and safely. They are used to transport food, packages, or goods. They are also used to scan different warehouse materials. They are preferred for transporting urgent or frequently sent small parcels and for delivery to hard-to-reach areas. Using drones to rapidly deliver small packages for logistics over short distances saves labor. Soon, drones may be an important player in the delivery options of packages. As long as there is a drone airport of tens of square meters, high-speed logistics can be achieved. One of the most important and pending problems in logistics use is that drones need a sufficient carrying capacity.

Some of these applications are displayed in Figure 1.11 [5].

Figure 1.11 Some of the applications of drones [5].

1.5 BENEFITS

Drones are used by a broad range of military forces, from Argentina to the US. They make it possible to collect data and access information in a variety of ways while saving time and minimizing costs. The commercial use of drones is transforming various industries and providing significant benefits, from increased efficiency and cost savings to improved safety and reduced environmental impact. Other benefits of drones include the following [9]:

- *Safety:* Drones can be safer than traditional methods, such as when inspecting high-rises or other dangerous areas. Drones can also be equipped with safety features like collision avoidance systems and geofencing. Drones can reduce hazardous tasks and improve safety in operations. For example, in construction, drones can complete surveying work faster and safer than traditional methods.
- *Efficiency:* One of the significant benefits of drones is their ability to reach and capture images and footage from angles and heights that would be impossible or impractical for a human operator. Drones can complete inspections and other tasks faster than traditional methods, which can increase productivity and reduce downtime. Reducing costs and expenditures while improving the efficiency of technological solutions is essential for large-scale industries.
- *Cost-effectiveness:* Drones can be more cost-efficient than other methods because you do not need to pay for labor or materials for people to come into your facility. They can reduce labor and operational costs. They can also be cheaper to buy and operate than other transportation methods, such as not requiring fuel.
- *Accessibility:* Drones can capture images and footage in remote or dangerous areas that were previously inaccessible. Drones can access remote or inaccessible areas that are difficult or dangerous for humans to reach.
- *Environmental Impact:* Drones can reduce waste and the volume of products like herbicides and water. They can also have zero impact on the soil, such as compaction and site disturbance.
- *Security:* Drones can provide security and surveillance services by capturing footage and monitoring remote locations.
- *Reduce Risk:* Industries such as mining, ports, gas, and large plants involve processes and applications where human power may be at risk or impossible to use.
- *Data Collection:* Drones can collect data faster and more efficiently than traditional methods. They can collect accurate data that can be used for a range of purposes, such as project updates, spotting for errors, streamlined communication, and decision-making.

1.7 CHALLENGES

Although drones have many benefits, there are some challenges to their widespread use. Such challenges include privacy and legal issues, regulatory challenges, potential security risks, obstacle detection, battery life, and hacking. Other challenges include the following [3].

- *Power Source:* One of the biggest problems with drone is the power requirement. Due to the limited flight times, developing drones capable of long flights with smaller and more powerful batteries remain the priority to solve this problem.

- *Security:* The greatest challenges of drones is security. All human-operated aircrafts inherently pose a risk of falling. Working on limited battery power, having fast spinning propellers, and the potential to fall from heights greatly threaten living things, structures, and the environment. The importance of safe use is increasing, especially with using drones in daily life. Law enforcement should be put in place to limit drones' interference with the privacy of others.
- *Hacking:* Drones are becoming a bigger target for cyberattacks as their use increases and their number increases. Hackers can intercept the transmitted data to take control of the drone. Extra measures should be developed to protect drones and the information they store.
- *Regulation:* The deployment of drones in manufacturing settings is gaining momentum. Adhering to local and federal regulations is becoming a pivotal concern for drone manufacturers. Responsible operations must comply with safety, security, and privacy regulations. It might be challenging, as legal issues are consistent and precise regulation can be complicated. European law, for example, regulates the activity of drones and a map by country of the areas where a drone can fly.
- *Weather:* Drones are sensitive to weather conditions, such as high winds and rain. High winds can make it difficult for drones to maintain a stable flight path, and rain can damage their batteries.
- *Collision Avoidance:* Ensuring that drones can reach their targets without colliding with obstacles is a challenge, especially when designing multi-drone systems.
- *Payload Capacity:* Drones have limited payload capacity, which can make it difficult to carry large sensors or equipment.
- *Noise:* Drones can be noisy, especially when flying over populated areas.
- *Cost:* Drones are expensive, and insurance companies may be reluctant to insure them.
- *Battery:* The brick-size batteries used by drones are heavy and get used up quickly. Gasoline engines are noisy and emit exhaust.

1.8 CONCLUSION

Drone technology is one of the advanced innovations in the realm of technology. As drone technology advances, engineers are pushing its boundaries to explore unconventional applications for drones. As the technology and equipment become more accessible, drones will be more active part of commercial and industrial operations. Being inexpensive and accessible, drone will be put to several novel uses around the world. Complementary technologies such as augmented reality and computer vision are likely to drive drone market growth and improve drone communication and intelligence. For more information about drones, the reader should consult the books in [10-20].

REFERENCES

[1] M. N. O. Sadiku, P. A. Adekunte, and J. O. Sadiku, "A primer on drones," *International Journal of Trend in Scientific Research and Development,* vol. 8, no. 4, July-August 2024, pp. 212-220.

[2] "How drones work and how to fly them," May 2024, https://dronelaunchacademy.com/resources/how-do-drones-work/

[3] K. M. Tugrul, "Drone technologies and applications," June 2023, https://www.intechopen.com/chapters/1154922

[4] T. Coleman, "The best drone 2024: Top flying cameras for all budgets," June 2024, https://www.techradar.com/news/best-drones

[5] N. Joshi, "10 stunning applications of drone technology," September 2017, https://www.allerin.com/blog/10-stunning-applications-of-drone-technology

[6] "Top 10 commercial uses for drones," April 2023, https://www.inspiredflight.com/news/top-10-commercial-uses-for-drones.php

[7] "What are the main applications of drones?" June 2024, https://www.jouav.com/blog/applications-of-drones.html

[8] "Transforming operations with drones: 10 Smart drone applications in manufacturing," March 2024, https://dac.digital/drone-applications-in-manufacturing/

[9] M. K. Hazarika, "Drones and its applications," https://events.development.asia/system/files/materials/2017/08/201708-drones-and-its-applications.pdf

[10] T. Westerlund and J. P. Queralta (eds.), *New Developments and Environmental Applications of Drones: Proceedings of FinDrones 2023*. Springer Nature Switzerland, 2024

[11] S. Sharma, *Drone Development from Concept to Flight: Design, Assemble, and Discover The Applications of Unmanned Aerial Vehicles*. Packt Publishing, 2024.

[12] A. Adams, *AI-Powered Drones: Applications and Challenges*. Independently Published, 2023.

[13] C. E. A. Ivar, *Entrenching Agricultural Drones Technology Guide: With this Simple Complete Beginners Manual Discover how Application of Drone Technology in Agriculture will Improve and Change the Growth of Agric −Bu*. Independently Published, 2024.

[14] D. Cvetkovic (ed.), *Drones - Various Applications*. Intechopen, 2024.

[15] M. LaFay, *Drones For Dummies*. Wiley, 2015.

[16] C. P. McCarthy, *Personal Drones*. ABDO Publishing Company, 2020.

[17] S. E. Kreps, *Drones: What Everyone Needs to Know*. Oxford University Press, 2016.

[18] D. R. Faust, *Police Drones*. Rosen Publishing Group, Incorporated, 2015.

[19] D. Hustad, *Discover Drones*. Lerner Publishing Group, 2016.

[20] D. R. Faust, *Military Drones*. PowerKids Press, 2015.

CHAPTER 2

DRONES IN AGRICULTURE

"For of all gainful professions, nothing is better, nothing more pleasing, nothing more delightful, nothing better becomes a well-bred man than agriculture."

- Marcus Tullius Cicero

2.1 INTRODUCTION

Agriculture represents the primary food source of the world. In this era of modern technologies, the structure of rural labor force worldwide has changed drastically. The agricultural industry is constantly evolving, and advancements in technology continue to reshape the way we approach farming and crop management. The global demand for agricultural machinery for production is growing. Today, farmers are dealing with increasingly complex issues such as climate change, water quality, soil quality, uncertain commodity prices, economic challenges in terms of productivity and cost-effectiveness, intense regulation, international competition, increasing labor cost, population increase, urbanization, an increasingly degraded environment, change in food preferences to name a few [1]. They are turning to high-tech to address the issues. They are compelled to seriously consider any tool that will boost productivity. One of such tools is drone, which is affordable and can easily be deployed. Agriculture significantly benefits from the commercialization of drone use. Aeronautical engineering and aerial imagery have evolved and were combined to give birth to drone technology.

A drone is a pilotless aircraft, designed to collect more accurate information than airplanes or satellites. Once the drone captures and processes the data, the data is sent to farmers in a readable format for management decisions. The data the drone collects will have to be processed with agriculture drone software [2]. The farmer can take the necessary actions to correct any problem.

The agriculture industry has embraced drones as indispensable tools for farmers around the world. Farmers can use drones for everything from pest control to plant health monitoring. Drones allow farmers to constantly monitor crop and livestock conditions by air. They are also helping farmers to optimize agriculture operations, increase productivity, increase crop production, monitor crop growth, reduce pollution, reduce wastage, and reduce time spent in the field [3].

This chapter provides an introduction on the use of drones in agriculture. It begins with defining drones. It explains the reason for using drones in agriculture. It covers different types of drones. It provides some applications of drones in agriculture. It highlights the benefits and challenges of using drones in agriculture. The last section concludes with comments.

2.2 WHAT ARE DRONES?

DRONE (Dynamic Remotely Operated Navigation Equipment) is commonly referred as unmanned aerial vehicle (UAV). Although drones were initially designed for military purposes, they are now widely applied in civilian settings such as agriculture, emergency, border patrol, disaster relief, and law enforcement. Drones are becoming popular for capturing aerial images, supplementing pre-existing imaging technology such as satellites and manned aircraft. They are used extensively in various commercial and industrial applications, ranging from the military cinematography, wedding videos, railway track monitoring, wildlife monitoring, delivery of small packages, security purposes, law enforcement operations, disaster management, and agriculture. The drones that are manufactured these days are getting smarter by integrating open source technology, smart sensors, better integration, more flight time, tracking down criminals, detecting forest, and other disaster areas [4].

Drones are equipped with all the software, sensors and hardware that a farmer will need to check the health of crop and survey farmland. A drone typically consists of propulsion and navigation systems, GPS, sensors, infrared cameras, software, and programmable controllers. The camera on a drone is a surveillance camera. Due to their size, drones cannot be boarded by a human body. They can be operated in two ways: directly by an operator and autonomously. Flying over the field, the drone takes high-resolution pictures with a camera. The drone is cost-effective approach to capture data about various crop conditions. A typical drone is shown in Figure 2.1 [5].

Figure 2.1 A typical drone [5].

2.3 WHY USE AGRICULTURE DRONES?

The drone used for agricultural activities is known as agriculture drone. There are two types of professionals who might want to own an agriculture drone: farmers and agriculture service providers.

With the world changing at fast pace, farmers will need to utilize new-generation technologies to address emerging challenges. Drones can they help farmers deal with a wide range of challenges. The use of drone technology in agriculture can become a game changer. By gaining access to a vast pool of data, farmers can increase crop yields, save time, reduce expenses, and act with accuracy and precision.

Drones can provide sustainable farming, improve yield, and increase farm productivity and profitability. They help farmers optimize the use of inputs such as seeds, fertilizers, water, and pesticides. Drone technology is used in crop scouting/monitoring, crop volume, generation of prescription maps, precision spraying, inspection of farm infrastructure (including irrigation), mapping and surveying of fields, crop damage assessment, and insurance claim forensics [6]. Agriculture drones are useful for aerial photography in livestock operations, spraying, drought assessment, monitoring, etc. Drone data is a powerful tool to help farmers visualize your fields. The high-resolution nature of drone data can be used by farmers to assess the fertility of crops, allowing agricultural professionals to more accurately apply fertilizer and reduce wastage.

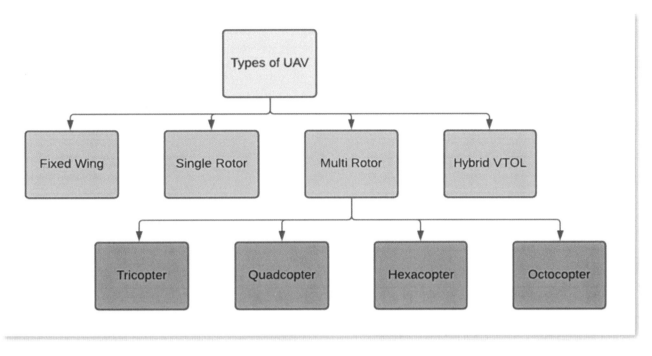

Figure 2.2 Types of drones or UAV [7].

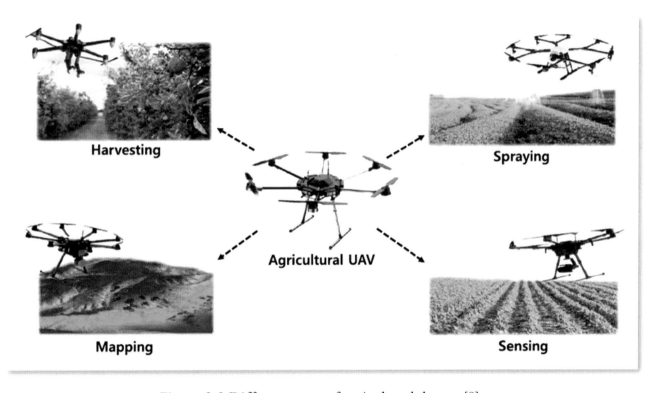

Figure 2.3 Different types of agricultural drones [8].

2.4 TYPES OF DRONES

Drones have emerged as powerful tools in modern agriculture, offering numerous benefits and opportunities for farmers. Use of drones in agriculture can vary widely. Farmers are finding new ways to use drones, such as for drone mapping and drone data analysis. Many agricultural drones are octocopters, hexacopters, and sometimes, quadcopters. The difference between these models is the payload capacity. There are several bases for categorizing agricultural drones. One way of classifying them is shown in Figure 2.2 [7]. Drones in agriculture have become a powerful tool for monitoring crops and livestock. Figure 2.3 shows another way to categorize agricultural drones, based on their various applications [8]. Drones generally fall into four distinct types:

- *Spraying Drones:* Spraying fertilizer and pesticides has been done by lightweight planes for over half a century. The ability of drones to move around quickly to their intended destinations is one of the top uses of drones in agriculture. Drones are ideal for monitoring and sensing practices as they can rapidly cover land to monitor the growth and health of crops and soil. Drones having this capability can spray insecticides and fertilizers on crops to nourish them and give them the nutrients they require. The right farming drones and spraying payloads can distribute chemicals evenly and efficiently. In addition to covering more land at a lower price, drone spraying can offer major environmental benefits. A typical spraying drone is displayed in Figure 2.4 [9].

Figure 2.4 A typical spraying drone [9].

- *Sensing Drones:* Drones are most commonly used as a platform to carry sensors to record observations about growing crops. This mission is no different than that of other platforms such as satellites and airplanes, which have historically been used for this purpose. Many types of sensors may be mounted on a drone. The sensor selection is based primarily on the end use goals. The remote-sensing capabilities of drones can provide farmers with crucial data to anticipate issues at an early stage and promptly make suitable interventions. A typical sensing drone is shown in Figure 2.5 [10]. The drone includes a downwelling light sensor (DLS) and an RGB (red, green, and blue) camera.

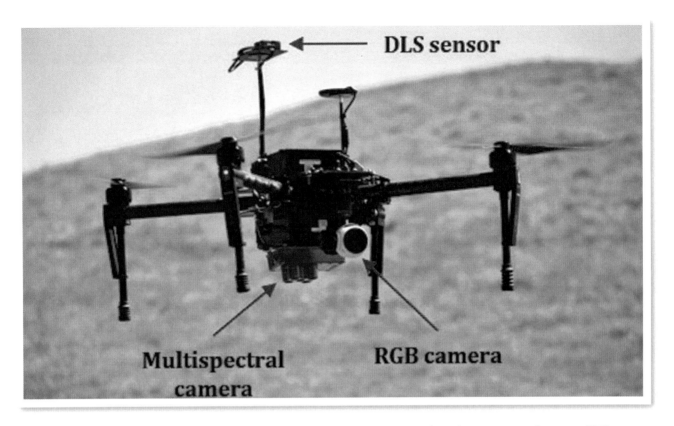

Figure 2.5 A typical sensing drone or a quadcopter equipped with camera and sensor [10].

- *Mapping Drones:* Drones are able to stay in the air for long periods of time and can cover large areas quickly. This makes them ideal for agricultural surveys, where they can be used to collect data about crop health and yield. Drones may be used to support and develop selected agriculture operations like mapping. Mapping and surveying drones are used to create high-resolution maps and 3D models of agricultural fields. They may be used to simply view fields from above, conduct systematic mapping missions, or map out an area and create new insights, taking the guesswork out of much of the growing process. By using drones for agriculture mapping, farmers can stay updated on the health of plants. Mapping drone technology can also be used for crop rotation and weeding operations.
- *Harvesting Drones*: Farmers are re-inventing forestry by harvesting from the air with high-capacity drones. Drones use artificial intelligence and machine learning to detect ripe apples before pulling the fruit from the tree. Companies in Israel have designed fruit-picking drones to make farms more efficient at a time of worker shortages and climate change. Fruit-picking drones the future of harvesting. A typical harvesting drone is displayed in Figure 2.6 [11].

Figure 2.6 A typical harvesting drone [11].

Figure 2.7 Applications of drones in agriculture [12].

2.5 APPLICATIONS

Agriculture has diverse and attractive uses for drones. Drone technology is currently being used for a variety of applications such as monitoring, mapping, irrigation, crop inspection, spraying, and surveying entire fields. Some common applications are illustrated in Figure 2.7 [12] and presented as follows [13]:

- *Monitoring:* Agricultural fields often occupy large areas, making surveillance task very difficult. A major challenge in farming is inefficient crop monitoring of vast fields. Drones are useful for real time monitoring large areas of farmland, in a more accurate and cost effective manner than satellite imagery, helping farmers make data-driven decisions. Farmers can cost-effectively monitor the health of crops and livestock using drones. Remotely accessing fields can help detect bacterial early. Drones equipped with the appropriate sensors can identify which parts of a field need more water. They can monitor any type of crop during its growing season. Parameters for monitoring include crop health, vegetation indices, plant height, plant scouting, water needs, and soil analysis.
- *Mapping:* The process of using a drone to map or survey crops is straightforward. The drone can survey the crops periodically: daily, hourly or weekly. Mapping with software gives agriculture customers detailed insights about their crops that would not be noticeable otherwise. By using drones for agriculture mapping, farmers can stay updated on the health of plants. Based on accurate, real-time information, farmers can take measures to improve the state of plants in any location. Drones could be used to produce accurate 3-D maps that can be used to conduct soil analysis on soil property, moisture content, and soil erosion.
- *Seed Planting:* Drone planting allows shooting seed pods into prepared soil. Farmers can use drones to deliver seeds, herbicides, fertilizer, and water. This minimizes the need for on-the-ground planting, which can be costly, labor intensive, and time consuming.
- *Spraying:* Drones can spray fields with water, fertilizers or herbicides, reducing costs, manual labor and time spent on these processes. This limits human contact with fertilizers, pesticides and other harmful chemicals. Drones can detect infected areas with sensors and cameras.
- *Precision Agriculture:* Precision agriculture, crop management that uses GPS and big data, refers to the approach farmers manage crops to ensure efficiency of inputs and to maximize productivity, quality, and yield. The use of drones for precision agriculture is gaining momentum because of their capability to deliver the most up-to-date info fast and efficiently. The use of GPS technology and GIS tools allow precision agriculture practices that enable monitoring and mapping of yield and crop parameter data within fields. Products can be traced from farm to fork using GPS locations for every point in the journey.
- *Sustainable Agriculture*: Sustainable land management can reverse the impact of climate change on land degradation. Agriculture can serve as an effective approach to sustainable agricultural management that allows agronomists, agricultural engineers, and farmers to help streamline their operations. The potential for drones in the improvement of sustainable agriculture is huge. Drones are helping to improve agriculture and achieve the Sustainable Development Goals.

Other uses of drones in agriculture include irrigation, seed planting, scheduling seeding and harvesting processes, and reducing usage of scarce resources.

2.6 BENEFITS

Just like any technology, drones have their pros and cons that farmers ought to know before investing in one. The evolution of drone technology and its overall affordability account for the increased application of drones. Having a drone in the sky allows you to find potential yield limiting problems in a timely fashion. Agricultural drones are known for their resource conservation and environmental friendliness. Drones provide real-time and accurate data that farmers can act on immediately. Drones could lift farmers out of poverty by providing invaluable data to make informed and prompt decisions while saving both time and money.

Other benefits of using drones for agriculture include the following [14-16]:

- *Increased Productivity:* Agricultural drones support the effort to meet the demand of feeding increasing world population. Images from drones can indicate the development of a crop with precision. Improving agricultural productivity is the key to building prosperity in smallholder communities and supporting their local economies.
- *Increased Accuracy:* Drones can help farmers to map their fields with great accuracy. Mapping drone technology allows farmers to more accurately identify areas of their fields that may need attention, such as dead grass or excessive weed growth.
- *Increased Efficiency:* Drone technology is often more efficient than using machinery or manual labor to collect this data. Drones can help farmers to save time and money by allowing them to quickly and easily assess the health of their crops.
- *Increased Crop Yields*: Drone maps of fields may increase yields by allowing growers to monitor their plants closely for signs of disease or infestation. Drone technology will help farmers spot issues on their crops before it's too late.
- *Adapting to Climate Change*: Extreme weather conditions are on the rise. Climate change continues to create new layers of complexity for the agriculture industry. It is having a major negative impact on food security. It is limiting productivity, such as drought, flood, and damage caused by storms. With the use of drone technology, farmers are successfully working towards sustainability.
- *Reduce Environmental Impact:* With the help of drones, farmers can cut down on agricultural runoff. Rather than spraying an entire field, which can lead to a negative environmental impact, agriculture drones for spraying can apply spot treatments of pesticides and fertilizers. Using drones for spot jobs in place of large full-field sprayers or crop dusters can also help reduce air-polluting emissions, as well as decrease input costs for farmers. Drone technology can also play a role in reducing CO_2 emissions from agricultural operations.
- *Multiplier Effect:* The drone technology generates secondary employment opportunities in rural areas, from drone operator to computer engineering positions.
- *Save Time:* Maintaining land, crops, and livestock is hands-on work, and that can take a lot of time. Drone technology provides agribusinesses with more in-depth results in real time. With images collected from a farming drone, agribusiness owners can get high-definition photos, videos, and data within minutes and save farmers time.

- *Cost Savings:* Drones can more easily and economically fly multiple times during the season. In addition to the time savings, high-resolution drone imagery offers more insights compared to more pixelated satellite images.
- *Radio-tracking:* Using drones saves time and money when radio-tracking animals.

We know that radio-tracking animals from the ground can be challenging. Searching for radio-signals across vast, difficult terrain like wetlands or rugged mountains takes a huge amount of time and effort, and the expenses certainly add up too. With drones, you can cost-effectively collect more radio-telemetry data, more often, and with less effort.

- *Help Troubleshooting:* Drones may also help troubleshoot issues. Agriculture small business owners can use an agriculture drone to monitor critical areas such as irrigation systems for potential leaks or damage before it becomes a significant issue, harming crops. A farming drone can help identify potentially dangerous chemicals or bacteria.
- *Improve Health:* Safety is often a primary concern for agribusiness, and agriculture drones have the potential to help improve this in critical ways. For example, if you suspect some of your livestock is sick or injured, drones can help you track their movement and potentially spot lethargic animals who may need help.
- *Security:* Drone cameras can provide an overview of farm operations throughout the day to ensure operations are running smoothly and to locate equipment being used. Security drones can be deployed to monitor fencing and perimeters of more valuable crops instead of employing more security personnel.

Figure 2.8 displays some of benefits of drones in agriculture. Additional benefits for agriculture drones include increase yields, easy to learn, save time, increased ROI, ease of use, integrated GIS mapping, color contrast crop health imaging, disease prevention, and more visibility.

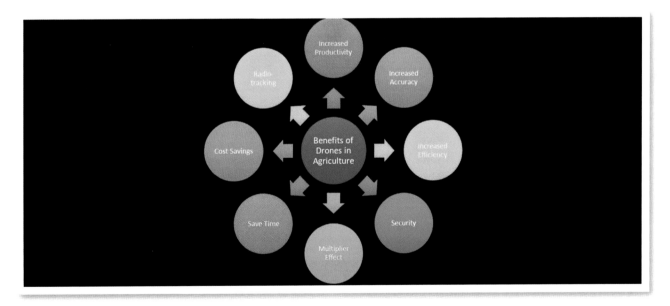

Figure 2.8 Some benefits of drones in agriculture.

2.7 CHALLENGES

The world-wide farming system faces tremendous challenges. There are some unsolved challenges that need attention for better implementation of UAVs such as battery efficiency, low flight time, communication distance and payload, pilot involvement, engine power, stability and reliability, sensors' quality due to payload weight limitations, implementation costs, and aviation regulation, etc. Barriers to widespread drone adoption include safety of drone operations, privacy issues, insurance-coverage questions, its complexity, and cost of using it. It is difficult to fly drones in extreme conditions. At the moment, drones cannot communicate directly with agricultural equipment. Other challenges include [17-19]:

- *High Costs:* Agriculture drones are generally expensive in the short run but worthwhile in the long run. The software and hardware that accompany drones could be expensive. High costs prohibit many farmers from using drone flight services. These include capital costs for drone equipment and labor costs for drone navigation.
- *Connectivity:* Operating drone will require some connectivity. For example, agriculture drone software often use Google Maps as their base layer. So it requires the Internet to work properly and lack of connectivity, particularly in remote areas, can be a challenge.
- *Limitations:* Strict regulations limit drones to only small areas in a single flight. There are also battery power limitations.
- *Flight Time and Flight Range:* Most drones have less flight time and covers less area. They have a short flight time of between 20 minutes to an hour. This makes limits the flight range. Drones that can offer longer flight time and longer range are relatively expensive.
- *Regulation:* Drones should be used responsibly. Farming with drones is considered commercial. It requires obtaining a remote pilot certificate from FAA. FAA requires drones fly at an altitude of not more than 400 feet. It is crucial for operators to keep up to date with regulations, such as individuals' privacy. Many countries like Canada, Malaysia, Singapore, and Australia have made laws regarding use of drones.
- *Drones Must be Registered:* If your drone weighs between 250g to 20kg, then you must register your drone at Federal Aviation Administration in the US. Anyone who will fly must pass a theory test and register to get an operator ID.
- *Special Skills:* Although drones can be used for entertainment, any professional use of drones in agriculture requires expertise. And, in the ideal case, only certified drone pilots with special skills can operate these unmanned aerial vehicles for in-field monitoring.
- *Drones Don't Fly Long:* Typically, an agriculture drone flies between 10 and 35 minutes. For big farms, this is not enough. However, longer flights are an attribute for more expensive drones.
- *Risks:* Drones are prone to interference with aircrafts since they share the same airspace. For the purpose of agriculture use, the drone must fly below 400 feet or 121 meters. There are risks in operating drones next to special areas like airports, aerodrome or other airfields. To avoid collisions, an agriculture drone should never fly near an airport or close to aircraft.

2.8 CONCLUSION

Drones, which were once a reserve of the military, are now redefining agriculture. Agricultural drones are unmanned aerial vehicles used to monitor the health of crops and livestock. They can transform modern farming in a number of ways. The number of farmers adopting drones in their farms is growing. Drone technology keeps improving daily. The agricultural community is just scratching the surface of what drone technology can provide to the industry. With many manufacturers producing drones, it is hoped that the cost of the drones and the accompanying equipment will reduce. The future of drones in agriculture appears promising. Agricultural drones are here to stay. More information about agricultural drones can be found in the books in [20- 33].

REFERENCES

[1] P. Daponte et al., "A review on the use of drones for precision agriculture," *Earth and Environmental Science*, vol. 275, 2019.

[2] A. Chalimov, "Why use agriculture drones? Main benefits and best practices," October 2019, https://easternpeak.com/blog/why-use-agriculture-drones-main-benefits-and-best-practices/#:~:text=As%20drones%20become%20an%20integral,a%20great%20number%20of%20benefits.&text=By%20gaining%20access%20to%20a,with%20unparallelled%20accuracy%20and%20precision.

[3] M. N. O. Sadiku, K. B. Olanrewaju, S. S. Adekunte, and S. M. Musa, "Drones in agriculture," *International Journal of Scientific & Engineering Research*, vol. 12, no. 9, September 2021, pp. 783-796.

[4] V. Puri, A. Nayyar, and L. Raja, "Agriculture drones: A modern breakthrough in precision agriculture," *Journal of Statistics and Management Systems*, vol. 20, no. 4, 2017, pp. 507-518.

[5] M. Choudhary, "What are popular uses of drones?" July 2019, https://www.geospatialworld.net/article/what-are-popular-uses-of-drones/ -

[6] *Drones on the Horizon Transforming Africa's Agriculture.* Gauteng, South Africa: NEPAD, 2017.

[7] P. Velusamy et al., "Unmanned aerial vehicles (UAV) in precision agriculture: applications and challenges," *Energies*, vol. 15, 2022/

[8] N. Islam et al., "A review of applications and communication technologies for Internet of things (IoT) and unmanned aerial vehicle (UAV) based sustainable smart farming," *Sustainability*, vol. 13, no. 4, February 2021.

[9] P. Kipkemoi, "The pros and cons of drones in agriculture," August 2023, https://www.droneguru.net/the-pros-and-cons-of-drones-in-agriculture/

[10] L. J. Thompson, Y. Shi, and R. B. Ferguson, "Getting started with drones in agriculture," December 2017, https://extensionpublications.unl.edu/assets/html/g2296/build/g2296.htm

[11] A. Ranti et al., "Drone: The green technology for future agriculture," *Harit Dhara,* vol. 2, no. 1, January – June 2019.

[12] J. Jeensen, "Agricultural drones: How drones are revolutionizing agriculture and how to break into this booming market," April 2019, https://uavcoach.com/agricultural-drones/#:~:text=Drones%20are%20transforming%20how%20agriculture,ll%20improve%20long%2Dterm%20success.

[13] J. Hayden, "Pros and cons of drones in agriculture [2019 update],"
https://drones-pro.com/pros-and-cons-of-drones-in-agriculture/

[14] Drone USA, "The benefits of drones to the agricultural industry," October 2018,
https://www.droneusainc.com/articles/the-benefits-of-drones-to-the-agricultural-industry

[15] "The benefits of drones in agribusiness,"
https://www.travelers.com/resources/business-industries/agribusiness/benefits-of-drones-in-agribusiness

[16] J. Leslie, "Benefits of using drones in agriculture,"
https://dronesurveyservices.com/benefits-of-using-drones-in-agriculture/

[17] P. Kipkemoi, "The pros and cons of drones in agriculture," August 2020,
https://www.droneguru.net/the-pros-and-cons-of-drones-in-agriculture/

[18] B. Pinguet, "The role of drone technology in sustainable agriculture," April 2020,
https://www.precisionag.com/in-field-technologies/drones-uavs/the-role-of-drone-technology-in-sustainable-agriculture/#:~:text=The%20use%20of%20drones%20in,effective%20insights%20into%20their%20crops.

[19] M. Kuzmenko, "Full list of disadvantages in agriculture drones," June 2020,
https://www.petiolepro.com/blog/disadvantages-of-agriculture-drones/

[20] K. R. Krishna, *Agricultural Drones: A Peaceful Pursuit.* Waretown, NJ: Apply Academic Press, 2018.

[21] K. R. Krishna, *Push Button Agriculture: Robotics, Drone, Satellite-Guided.* Waretown, NJ: Apply Academic Press, 2016.

[22] S. Rose, *Agricultural Drones.* Capstone Press, 2017.

[23] A. Ingole, S. Giri, and B. Tone, *Advance Agricultural Drone System: Advance Agriculture Drone For Collecting Data and Spraying of Pesticides and Fertilizer.* LAP LAMBERT Academic Publishing, 2019.

[24] G. Sylvester (ed.), *E-Agriculture in Action: Drones for Agriculture.* Bangkok: FAO and ITU, 2018.

[25] D. Soesilo and G. Rambaldi, *Drones in Agriculture in Africa and Other ACP Countries: A Survey On Perceptions and Applications.* CTA, 2017.

[26] Food and Agriculture Organization of the United Nations, *E-agriculture in Action: Drones for Agriculture.* Food & Agriculture Org., 2018.

[27] L. Jupp, *Precision Farming From Above: How Commercial Drone Systems are Helping Farmers Improve Crop Management, Increase Crop Yields and Create More Profitable Farms.* Writing Matters Publishing, 2018.

[28] K. R. Krishna, *Unmanned Aerial Vehicle Systems in Crop Production: A Compendium.* Apple Academic Press, 2019.

[29] A. Diantini, M. De Marchi, and S. E. Pappalardo, *Drones and Geographical Information Technologies in Agroecology and Organic Farming: Contributions to Technological Sovereignty.* CRC Press, 2021.

[30] L. J. Thompson, Y. Shi, and R. B. Ferguson, *Getting Started with Drones in Agriculture.* University of Nebraska-Lincoln, Extension, 2017.

[31] K. S. Subramanian, *Drones in Digital Agriculture.* Astral International Pvt. Limited, 2022.

[32] R. Avtar and T. Watanabe, *Unmanned Aerial Vehicle: Applications in Agriculture and Environment.* Springer, 2019.

[33] N. Narra, P. Linna, and T. Lipping (eds.), *New Developments and Environmental Applications of Drones: Proceedings of FinDrones.* Springer 2020.

CHAPTER 3

DRONES IN BUSINESS

"Every industry and every organization will have to transform itself in the next few years. What is coming at us is bigger than the original internet, and you need to understand it, get on board with it, and figure out how to transform your business."

— Tim O'Reilly

3.1 INTRODUCTION

Technology has had a huge influence on workplace productivity. Drones are no exception. Drones provide aerial data that can help automate many tedious tasks.

They play a part in the digital transformation. There is no denying that drones are hot technology topics these days, with many people eager to get their hands on one for personal or commercial use.

A drone, also known as an Unmanned Aerial Vehicle (UAV), is an aircraft without a human pilot aboard. It may be regarded as a small aircraft that can fly without a human pilot, usually made of lightweight materials, that can be remotely controlled or fly autonomously. Drones are robots typically remotely controlled by a pilot and with the ability to fly or move. It seems drones are breaking into every industry. Drones can serve as a platform for different applications and business models. In many business activities, drones can replace traditional methods of operation.

Drones are quickly gaining traction among both individuals and major companies. Drones can vary in shape and size, but the main core elements (battery, microcontroller, motor, sensors) essentially remain the same. With less human operation and no safety infrastructure, drones can reduce time and costs. Drones are being used to gather information for businesses and support commerce. They are commonly used in healthcare, agriculture, military, insurance, delivery, etc. [1,2].

It seems drones are breaking into every industry. Business and commerce are no exception to the drone craze. Drones are being used in collecting aerial footage for film and television and inspecting hard-to-reach areas—such as power lines, pipelines, and transmission infrastructure—for utility and energy companies. Businesses such as fast food restaurants, theme parks, schools, and governments use drones to conduct aerial traffic studies. Business executives are discovering ways to use drones to gather intelligence on the competition. Drones are already being applied by innovative industry leaders. For example, Amazon's Prime Air fleet drops off packages at your doorstep [3].

This chapter introduces readers to drones and their various applications in the business world. It begins by describing what drones are. It provides some applications of drones in business. It highlights the benefits and challenges of drones in business. It covers the global uses of drones in business. The last section concludes with comments.

3.2 WHAT ARE DRONES?

Drones are autonomous robots that fly in the sky. They may also be regarded as a pilotless aircraft that were initially used by the military but are now used for scientific and commercial purposes. The word "drone" was coined due to the similarity of its sound to a male bee. Drones are pilotless aircraft formally known as unmanned aerial vehicles (UAVs) or unmanned aircraft systems (UAS). Drones are also called "remotely piloted vehicles" or "unmanned aerial systems." Drones were first used in the 1990s by military organizations. The notion of drones began around 1918 when the US Navy commissioned militarized unmanned aerial vehicle (UAV) built by Charles Kettering. Their original use was to take strategic pictures for the military. From the beginning of the 21st Century, civil activities started to get more attention.

Drones are classified in different ways, according to their size, weight, flight time, commercial or military, and cost. The US Federal Aviation Administration (FAA) defines consumer and commercial drones as those that weigh less than1.0 lb. (0.45kg) with approximately a maximum 500 m altitude and 2km range from the base operator.

Drones have been used for blood, food, and package delivery. They are now used in different fields, including transportation, healthcare, news media, commerce, safety and security, disaster management, rescue operations, crop monitoring, weather tracking, environmental protection, intelligence gathering, surveillance, aerial photography, express shipping, recreation, agriculture, wildlife, military, law enforcement, home, cemetery management, power, infrastructure, telecom sectors, marine, weather forecasting, sports, space, insurance, hotels, journalism/news coverage, and logistics [4,5]. Drones have been used by the military in combat and for humanitarian aid. Drones have emerged as interaction devices in home and research applications. A drone can be used as a companion, personal drone, agent, sensing tool, delivery tool, ambulance drone, etc. Drones are commonly used by hobbyists just for the fun of it. Thus, we have different kinds of drones: military or armed drones, healthcare drones, medical drones, biomedical drones, smart drones, humanitarian drones, collaborative drones, ambulance drones, courier drones, nano or micro or mini drones, etc.

The technology involved in drone construction is impressive. Modern drones are empowered by sensor technology. Accelerometers are often used to determine the position and orientation of the drone in flight. Inertial measurement units and global positioning systems (GPS) are critical for maintaining direction and flight paths [6]. Drones provide a reliable connection regarding tracking and remote control purposes and communicate these with users. They can maneuver unobtrusively above the ground towards a target user without disturbing human movement on the ground.

3.3 APPLICATIONS

Drones play major roles in everyday life, from package delivery to sporting events. Many industries have found ways to use drones to benefit their business. Drones continue to grow in popularity as entrepreneurs discover more and more applications for them. Industries that commonly use drones for business include the following [7-9]:

- *Agriculture*: Drones can be an exciting means of merging technology with agriculture. Many countries are large exporters of food. The high demand for food allows drones to be used for various tasks. Some drones can deliver water, pesticides, or herbicides. Drones can assist both large and small farming operations with disease management services. Drones carrying remote sensors can collect data points to provide precise evaluations of, for instance, soil moisture or crop yields. Farmers can use drones to get a "big picture" idea of the crops their competitors are growing. Some drones provide farmers with crucial information about the weather on how crops are doing. Figure 3.1 shows how drones are used in agriculture [10].

Figure 3.1 How drones are used in agriculture [10].

- *Restaurants:* Fast food chains and other restaurants use drones for aerial traffic studies. For example, Chick-fil-A has hired drone pilots to perform aerial traffic studies. With a drone, the pilot can hold the drone in a static location above the property and record during peak hours, allowing management to determine where the pile-ups happen. Chick-fil-A performs an aerial traffic study of customers in a queue, as shown in Figure 3.2 [11]. The Dominos delivery drone is still very much in the test phase.

Figure 3.2 Drones are used for aerial traffic studies [11].

Figure 3.3 A real estate drone [12].

- *Real Estate:* Many real estate professionals now use drones to help with their marketing efforts. Most drones have cameras that give real estate companies an opportunity to capture photos and videos of their properties. You can use a drone in advertising to offer your customers innovative campaign options. Most homebuyers begin their search online, where they expect to find photos and videos of different homes. A real estate drone is shown in Figure 3.3 [12].
- *Insurance:* Insurance companies were among the first to adopt drones. Many insurers are leaping into technological advances by using drones in different ways. Drones can be used to gather data before a risk is insured and to assess damage after an event. They have proved to be useful for insurance companies after events like natural disasters.
- *Construction:* Drones are being used in various ways and are changing the way the construction industry works. They have helped revolutionize the entire construction project life cycle from beginning to end. Construction companies use drones to help keep clients up-to-date on the progress of their work. Drones are used to reduce labor times for surveying land, improving infrastructure, providing more accurate surveillance, and adding efficiency to inspections. Drones are also used in the inspections of bridges. A typical use of construction in construction is shown in Figure 3.4 [13]

Figure 3.4 A typical use of drone in construction [13].

- *Photography:* Most businesses use drones for video and photography. Drone photography is the ability to capture both still images and video by using a drone. Drones ensure wedding couples have interesting and different photographic shots from various angles and heights. Drones can

both fly in the sky and dive into the ocean. One profitable way to make money with drones is to become a wedding photographer and videographer.
- *Package Delivery:* Drones are being used for package delivery. Amazon, Walmart, and UPS are currently delivering packages by using drones. Traditional postal service providers such as USPS, UPS, and FedEx have been using drones for shipping and handling for major online retailers. They are looking to save money with drones. Package delivery is usually done in rural and remote areas. Amazon Prime Air is a drone delivery system that anticipates package deliveries in 30 minutes or less. A typical package delivery is shown in Figure 3.5 [8].

Figure 3.5 A typical package delivery [8].

- *Security:* Drones and robotics are among the most exciting and transformational emerging technologies in the security industry, and they are solutions that practitioners are increasingly eager to deploy.
- *Utility Companies:* Power and utility companies are using drones to inspect pipelines, powerlines, and similar infrastructure, leveraging the ease of drone image captures.

Other applications include aerial surveying, sports video, drone videography, drone advertising and marketing, filmmaking, maintenance, disaster relief, safety inspection, data monitoring, police and fire departments, military, security surveillance services, telecom industry, entertainment industry, warehousing and inventory industry, engineering industry, underwater inspections, mapping, and surveying, etc. These various applications have allowed drone pilots and manufacturers to capitalize on the new technology.

3.4 BENEFITS

Drones have gained popularity in recent years for both business and personal use.

They have recently become widely used for both recreational and commercial purposes. Drones are a versatile technology that many companies have started to use. Drones are saving companies time and money. The economic impacts associated with drone integration include job creation and billion-dollar growth. Changing customer expectations and increasing competition conditions have made it necessary to develop applications that include modern technologies, such as drones, in logistics processes.

Drones are starting to help make many tasks faster and life easier. Other benefits of drones in business include the following [15,16].

- *Disaster Relief:* Disaster relief organizations now have new ways to get relief packages into the hands of those who need them, especially in areas that are not yet safe for human travel. Drones have been used in emergency situations to locate missing people. With drones, pilots can safely monitor the damage, assess the terrain, and rescue people quickly.
- *Safety Inspections:* Safety is crucial to industrial worksites, and drones can reduce the cost while multiplying the accuracy of inspections. Remote inspections use drones that can maneuver into hard-to-reach areas to gather information about the safety of structures such as bridges, buildings, dams, and other construction sites.
- *Drone Delivery:* Companies such as Amazon and UPS have experimented with delivering goods with drones as a faster, cheaper alternative to other delivery systems, especially in rural areas. The drones can fly to and from their destinations, requiring no pilot control and no fuel cost.
- *Data Gathering:* Rather than gathering information from the ground, drones are capable of setting specific flight paths and gathering accurate information from the air, and reporting back precise data.
- *Asset management:* Currently, the most compelling drone application for multiple industry verticals is asset management and protection. Whether the assets are power lines, buildings, humans, wildlife, or roads, drones are deployed for rapid, efficient inventory and survey purposes.
- *Maintenance:* Drones support ongoing routine facility inspections at lower cost and risk, particularly in potentially hazardous areas such as power lines or power plants or in the case of very tall structures such as radio antennas or bridges. Drones also allow for wear and damage assessments. They can also be used for routine deliveries of consumables and spares between pre-programmed launch and landing pads, thereby reducing cost and potential downtime.
- *Mapping:* Drones have had a large impact on the data mapping industry in recent years and will continue to change the way we survey and map land. Drones' mapping applications offer an efficient geographic survey tool for developers, civil engineers, and local authorities. But the most forward-thinking are already using AI and GIS to analyze drone images and identify areas that need servicing.
- *Film and Multimedia*: Already widely used by film and photography professionals, the potential for drones to add new dimensions to multimedia is significant. Aerial visuals, now possible through drones, also offer opportunities to the real estate, hospitality, and tourism sectors.

- *Monitoring Construction:* Construction companies now use drone services to help keep clients up-to-date on the progress of their work. Construction sites are often massive by nature and are always changing. Construction firms can now hire drone operators to visit a construction site at regularly scheduled intervals and quickly turn around imagery to share with clients. Drones offer construction firms and developers efficient new 2D and 3D mapping methods. This feature allows for greater efficiencies from pre-construction through to the maintenance phase.
- *Drone Surveying:* This is also known as aerial surveying. It provides a faster, safer, and cost-efficient way to survey at heights. They enable construction engineers to make vital decisions. The use of drones for mapping and geographic surveys can also add significant value to marketers and brands researching target areas. Unlike satellite imaging, traditional aerial surveys, or human resources on the ground, drones can conduct highly targeted surveillance with live feeds to headquarters and no risk to human life. Engineers prefer aerial surveys for bridge inspection, roof inspection, and large building construction. By using drones, engineers reduce risk and enhance safety while working at high-construction sites.

Some of these benefits are depicted in Figure 3.6.

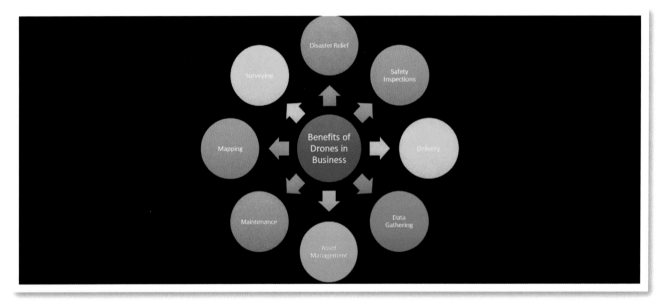

Figure 3.6 Some benefits of drones in business.

3.5 CHALLENGES

Drones, with their tremendous opportunities, come with a lot of challenges. Some have raised concerns that drones may still be used to infringe upon individuals' privacy rights. Some are outraged by the military's use of drones and fear terrorist attacks using drones as weapons. Rushing to implement drone technology without adequate preparation can lead to legal and financial disaster. Perhaps the most important challenges to implementing drone logistics are external terrorist attacks, the deficiencies of legal regulations for drone flights, the difficulty of flight permit processes, flight area limitations due to restrictive visual line of sight, and personal data protection. Therefore, it is necessary to provide technical infrastructure and flight arrangements to ensure drone logistics' applicability [16]. Other challenges include the following [17,18].

- *Bad Reputation:* Drones have got a bad reputation. Most people associate them with expensive military aircraft or small consumer toys. While the financial implications of drone use are robust and obvious, numerous consumers, states, and regulators believe sanctioned UAV use to be detrimental. All it takes is just one person using a drone wrongly or with bad intentions to significantly harm public safety.
- *Privacy:* This is a major concern for using drones for data collection. Without human monitoring, a drone is unable to ensure seamless deliveries. As drones can capture images, audio, and video from a distance, many have expressed concern that their use may violate the privacy of individuals. The public can view data gathering as intrusive, almost as if they are being spied on. These concerns over personal privacy have prompted federal legislation to be introduced.
- *Security:* This is always an issue for any high-tech products, and the drone is no exception. Drone communications with remote controllers are often encrypted. Small commercial drones often rely on existing encrypted communication methods, which are relatively easy to hijack. It is possible that other sources may attempt to jam or hack the drone's signals. The fear that sensitive data may end up in the wrong hands can create anxiety among the public.
- *Safety:* When drones are misused, safety is always the biggest concern. Drones are small, fast, and relatively difficult to detect compared to flying jets. This attribute is precisely the reason why there is a need for an identification system for drones and their operators. Drones can interfere with the flight patterns of other aircraft and pose potential safety threats. Terrorists are using drones to organize crime.
- *Regulation:* Aviation authorities such as the Federal Aviation Administration (FAA) in the US have started introducing several new regulations as more people own drones and also to give guidelines on how to operate drones safely and effectively. The drone pilots/operators must also register with the FAA before legally flying the drones.
- *Power:* Drones have limited payload and flight endurance. Adding extra sources of power or mechanisms to improve their payload capacity adds to drones' weight and manufacturing cost. Increasing the payload also reduces the flight time of the machine. While advancements in battery and engine technologies are crucial for longer flight times, the weight of unmanned aerial vehicles is also an important consideration.

3.7 GLOBAL USES OF DRONES IN BUSINESS

Drones have become widely used for both recreational and commercial purposes worldwide. For example, in France, the number of approved drone operators has risen from 86 in 2012 to 431 in 2013. In the US, drones are used to count sea lions in Alaska, monitor drug trafficking across borders, and conduct weather and environmental research. American and European drone technology is used to influence aviation majorly, but now China is influencing it more than ever because of its innovation in this field.

We consider how drones are employed in different regions of the world.

- *United States:* In the US, drones have quickly become majorly invested in tools for tech and retail giants like Amazon, Facebook, Wal-Mart, and Google, not to mention the various industries like real estate, police and fire departments, farming, cinema, construction, concerts, sporting events, and photography that stand to benefit greatly from commercial drone use [19]. Although the economic impact of drones is robust, the Federal Aviation Administration's (FAA) regulations, in conjunction with privacy and safety concerns, have delayed the full operation of commercial drone services. Thus, FAA regulations are the biggest obstacle to commercial drone usage. Companies like Amazon, Walmart, and Facebook face regulation and approval challenges from the US FAA. Flying drones commercially in the US requires passing a test and obtaining a Remote Pilot Certificate from the FAA. After getting the certificate, you must register your drones with the FAA. For drones operating in urban areas, drone insurance is a must. Once you become a drone pilot, you will get requests for all kinds of different drone jobs, such as real estate, construction, photography, advertising, land mapping, etc. [20].
- *United Kingdom*: The UK has taken a proactive approach to regulating drone use. It is embracing the use of drones, with applications ranging from the military to leisure activities. The potential of drones is apparent in the agricultural, healthcare, infrastructure, energy, logistics, and security industries. For example, the UK police forces use drones to help with searches, crowd control, and surveillance. The Civil Aviation Authority (CAA) has issued a set of regulations and guidelines to ensure the safe and responsible operation of drones. All drones weighing 250 grams or more must be registered with the CAA, and the pilot must be at least 18 years of age and have completed an approved safety test. Under the Drone Code, drones must be kept within 400 feet in altitude and 500 meters in distance from people and property. According to the CAA safety guidelines, drones must avoid flying in bad weather, drones must avoid other aircraft, and operators must never fly while under the influence of alcohol or drugs [21].
- *Europe:* Europe regards the emergence of civil applications of drones as a source of economic growth and jobs. The European Parliament has initiated the discussion and urges the active promotion of transparency, accountability, and the rule of law concerning the use of drones. The European Drone Strategy 2.0, adopted by the European Commission, sets out a vision for further developing the European drone market. It is possible to automatically program these flying machines by using and adapting open-source software such as MultiWii5, Pixhawk6, OpenPilot7, Dronecode8, or DIYdrones. One could imagine a drone controlled by an iPad or a smartphone and produced using 3D printing. Recently, European companies have increased the use of drones to provide various business services. Currently, Europe has over 1000 drone operators [22]. The changes in drone regulations made by the European Union Aviation Safety Agency (EASA) support the deployment of drones in various sectors such as infrastructure, agriculture, transport, entertainment, and security.
- *China:* This nation is the second-largest drone market in the world and is expected to close the gap with the leading market, the USA, in 2023. China's drone industry is thriving and profitable. China has already risen to one of the leading global manufacturers due to the country's rapid development. As the home country of DJI, the world's largest drone maker based in Shenzhen, drones are becoming increasingly popular in China. DJI has long been the global leader in drone manufacturing, holding over 70% of the global drone market share. With great support from the government, China is dominating the global consumer and commercial drone market. The Chinese market is currently the second-largest drone market in the world

and will continue closing the gap with the leading market, the United States. While drones were originally invented for military purposes, its their usage has been spread across various industries, including infrastructure development, construction, and transportation [23]. Figure 3.7 provides the top reasons China is one of the global leaders in the drone industry [24].

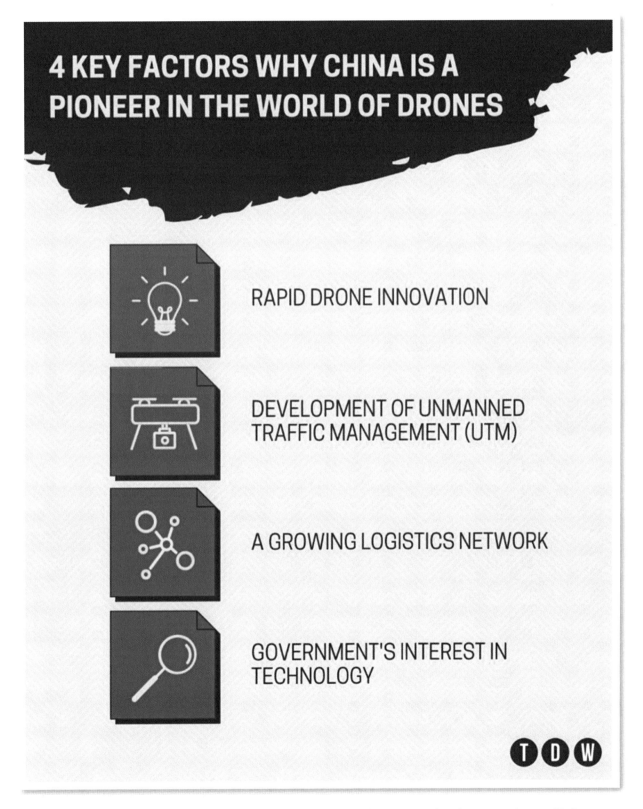

Figure 3.7 Top reasons China is one of the global leaders in the drones industry [24].

- *India:* India is gradually discovering the usage of drones in various sectors. Initially used in defense, photography, and recording videos, drones are now being engaged in food delivery, surveillance, geographical mapping, disaster management, search and rescue, and more. Currently, India accounts for 22.5% of total global drone imports. By 2025, India is forecasted to be the world's third-largest drone market. The government of India is planning to make India a Global Drone Hub by 2030, and a total of 12 central ministries are involved in trying to boost indigenous demand for drone services. Various initiatives like waivers for pilot permits, reduced and simplified procedures, the creation of new drone corridors, and incentives for local manufacturers are likely to allow drones to transform the scenario across numerous industries in the country. Drone-based applications are portrayed in Figure 3.8 [25].

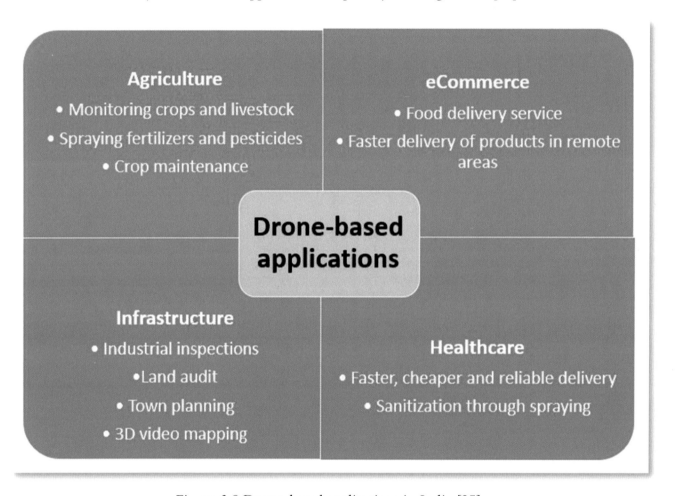

Figure 3.8 Drone-based applications in India [25].

- *Israel:* This is the most advanced state in the world when it comes to practically examining the anticipated architecture of control in unmanned airspace. Israel's population of 9.3 million people is largely packed in urban centers, with major cities like Tel Aviv and Jerusalem. While at the forefront of military and commercial drone technology, Israel's airspace is still limited with strict control, restricted commercial flying zones, and no flying zones along its border. Even though there are many successful Israeli drone companies, the ecosystem is still tilted toward security applications due to the lack of economic incentives to test and develop civil applications. Drones can assist in two ways – through monitoring and analysis and the operational aspect of creating new transportation lanes that will ease urban congestion. With every successful demonstration, Israel not only pushes for the ease of drone regulations and their incorporation in our day-to-day lives but also opens opportunities for every company worldwide to further develop and expand their products and services [26].

3.8 CONCLUSION

Drones, or Unmanned Aerial Vehicles (UAV), are aircraft without passengers or pilots that can fly throughout the airspace autonomously or by being controlled remotely from the ground. They are essentially aerial robots. From the initial unmanned "balloons" to the sophisticated models we have today, drone technology has come a long way. If you are interested in drones, be sure to consider the available options. To fly a drone professionally means that you can legally operate a drone for business purposes and get paid for the service. Figure 3.9 shows that drones will operate at different altitudes depending on their function [21].

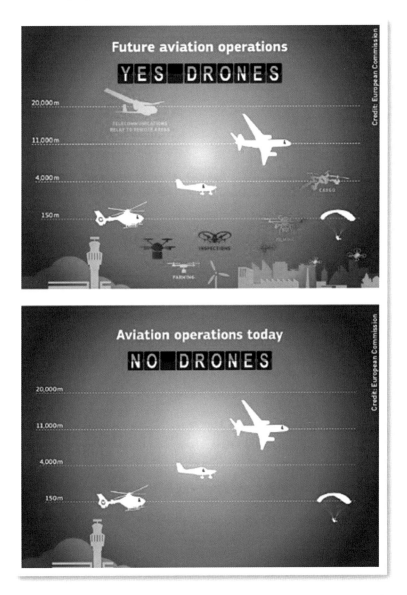

Figure 3.9 Drones will operate at many different altitudes depending upon their function [21].

Many business executives are quick to realize the myriad of benefits drones could offer by integrating technology into their operations. The business of drones is booming. The future of drones being used commercially by businesses has already begun. Drones are here to stay and will soon become mainstream. More information about drones in business can be found in the books in [27- 39].

REFERENCES

[1] M. N. O. Sadiku, O. D. Olaleye, P. Oyekanmi, and S. M. Musa, "Drones in healthcare: A primer," *International Journal of Trend in Research and Development*, vol. 8, no. 1, Jan.-Feb. 2021, pp. 39-41.

[2] M. N. O. Sadiku, K. B. Olanrewaju, S. S. Adekunte, and S. M. Musa, "Drones in agriculture," *International Journal of Scientific & Engineering Research,* vol. 12, no. 9, September 2021, pp. 783-796.

[3] M. N. O. Sadiku, U. C. Chukwu, and J. O. Sadiku, "Drones in business," *American Journal of Economics and Business Management*, vol. 6, no. 8, August 2023, pp. 43-48.

[4] S. H. Alsamh et al., "Survey on collaborative smart drones and Internet of things for improving smartness of smart cities," *IEEE Access,* vol. 7, 2019, pp. 128125- 128152.

[5] "38 Ways drones will impact society: From fighting war to forecasting weather, UAVs change everything," January 2020, https://www.cbinsights.com/research/drone-impact-society-uav/

[6] C. Winkler, "Sensor solutions play critical roles in enabling innovation in drone," June 2016, https://www.designworldonline.com/sensor-solutions-play-critical-roles-in-enabling-innovation-in-drones/

[7] K. Martinez, "5 industries using drones to benefit their business," https://www.rob-harris.com/5-industries-using-drones-to-benefit-their-business/

[8] "9 Ways drones are transforming business," September 2020, https://www.sme-news.co.uk/9-ways-drones-are-transforming-business/

[9] B. Cross, "The three business benefits of drones," https://www.forbes.com/sites/esri/2020/11/10/the-three-business-benefits-of-drones/?sh=5ed7b9862bf4

[10] "How to get into the drone business: Ideas, plans, models and business opportunities in 2021," Unknown Source.

[11] "Benefits of drone technology in business and commerce," https://www.droneblog.com/benefits-of-drone-technology-in-business-and-commerce/

[12] " What's the best drone for real estate photography?" June 2020, https://www.halfchrome.com/best-real-estate-drone/

[13] "Drone use in construction," https://flyguys.com/uav-industries/construction/

[14] "10 benefits of using drones for business," Jube 2017, https://www.bizcommunity.com/Article/196/706/163658.html

[15] "9 Ways drones are transforming business," https://www.dronegenuity.com/drones-transforming-business/

[16] S. Çıkmak, G. Kırbaç, and B. Kesici, "Analyzing the challenges to adoption of drones in the logistics sector using the best-worst method," *Business and Economics Research Journal,* vol. 14, no.2, 2023, pp. 227-242.

[17] "The three key challenges facing today's drone industry," September 2021, https://www.swidch.com/blogs/three-key-challenges-facing-the-drone-industry

[18] "4 of the biggest challenges in unmanned aerial vehicles," January 2020, https://www.switzermfg.com/biggest-challenges-unmanned-aerial-vehicles/

[19] "The future of drones in business & commerce," https://mondo.com/insights/future-of-drones-in-business-commerce/#:~:text=Not%20only%20are%20they%20already,%2C%20sporting%20events%2C%20and%20more.

[20] C. Yeager, "Establishing a drone business in 2022," December 2021, https://www.shutterstock.com/blog/drone-business-essentials?ds_eid=700000001400310&gclsrc=aw.ds&gclid=EAIaIQobChMIt_nIyNvX_gIV5MfjBx02QguuEAAYASAAEgLQr_D_BwE&kw=&utm_medium=cpc&ds_agid=58700005537898425&ds_ag=FF%3DDSA%20-%20Blog_AU%3DProspecting&ds_cid=71700000027388020&utm_source=GOOGLE&utm_campaign=CO%3DUS_LG%3DEN_BU%3DIMG_AD%3DDSA_TS%3Dlggeneric_RG%3DAMER_AB%3DACQ_CH%3DSEM_OG%3DCONV_PB%3DGoogle

[21] M. Frąckiewicz, "The use of drones in the United Kingdom: Aplications and regulations," February 2023, https://ts2.space/en/the-use-of-drones-in-the-united-kingdom-applications-and-regulations/

[22] "Drone regulation in the European Union amid a thriving market," https://room.eu.com/article/Drone_regulation_in_the_European_Union_amid_a_thriving_market

[23] "The rise of China's drone industry," https://eac-consulting.de/the-rise-of-chinas-drone-industry/

[24] B. Kitanovic, "Drone industry in China: A pioneer in the world of drones," May 2021, https://thedronesworld.net/drone-industry-in-china/

[25] A. Sharma, "India's drone industry: A flight to transformation," January 2023, https://www.tpci.in/indiabusinesstrade/blogs/indias-drone-industry-a-flight-to-transformation/

[26] J. Antunes, "The impact of drones in Israel for the enterprise," June 2022, https://www.commercialuavnews.com/drone-delivery/the-impact-of-drones-in-israel-for-the-enterprise

[27] D. Sumner and M. Farrell, *Launching Your Successful Drone Business*. Amazon Digital Services LLC, 2023.

[28] D. Rodwelll, *The $100,000 Drone Business*. Kingston Publishing, 2020.

[29] B. Halliday, *Drones: How to Make Money*. Amazon Digital Services LLC, 2019.

[30] J. LeMieux, *Uav and Drone Entrepreneurship*. Unmanned Vehicle University, 2013.

[31] P. R. Shaub, *Make Money with Drones Learn the Steps to Starting Your Own Drone Based Business*. CreateSpace Independent Publishing Platform, 2016.

[32] D. Preznuk, *The Drone Age: A Primer for Individuals and the Enterprise*. MC&W Publishing, 2016.

[33] S. C. Xercavins, *Business Plan to Implement One-hour On-demand Delivery Service by Drones*. Univeersitat Politecnica de Catalinya, 2016.

[34] D. Gonzalez, *+101 Ways to Make Money with Drones: Practical Suggestions for Starting a Business of Your Own and Profit Your Special Skills*. Amazon Digital Services LLC, 2019.

[35] J. Deans, *Become a U.S. Commercial Drone Pilot (Business Series)*. Self-Counsel Press, 2nd ed., 2016)

[36] S. Kaiser, *The Drone Business Guide Markets*. Amazon Digital Services, 2018.

[37] J. LeMieux, *Drones/UAVs Entrepreneurship:30 Businesses You Can Start.* CreateSpace Independent Publishing Platform, 2013.

[38] A. Wilson, *Mastering Drones: A Beginner's Guide To Start Making Money With Drones.* Adidas Wilson, undated.

[39] R. Blomquist and T. Larsen, *Six-Word Lessons for a Trustworthy Drone Business: 100 Lessons to Optimize the Drone Entrepreneur/Client Relationship.* Amazon Digital Services LLC, 2018.

CHAPTER 4

HEALTHCARE DRONES

"It is not the strongest of the species that survives, nor the most intelligent, but the one most responsive to change."

— Charles Darwin

4.1 INTRODUCTION

Technological advances are changing the world around us. They have revolutionized the medical field and changed the way healthcare is delivered. Drones (otherwise known as Unmanned aerial vehicles) are the next wave of technological advance that can make a huge impact on healthcare. They are gradually becoming a recognizable facet of everyday life.

Drones are autonomous or remotely controlled multipurpose aerial vehicles driven by aerodynamic forces. They are devices which are capable of sustained flight and do not need a human on board. Typically, a drone consists of an air frame, propulsion system, communication system, and navigation system [1]. Common drone configurations include fixed-wing, rotary-wing, multirotor, and hybrid designs. Their obvious usage areas are transportation and package delivery. As illustrated in Figure 4.1, compared with other transport modes, drones are the fastest and the least expensive [2]. Since a drone can fly over an inaccessible road, organizations have begun to use drones for healthcare delivery. While governments and regulators may be cautious about allowing drones to roam the skies, healthcare deliveries have a compelling reason to go for it. Amazon announced its plan to use drones to deliver packages to customers.

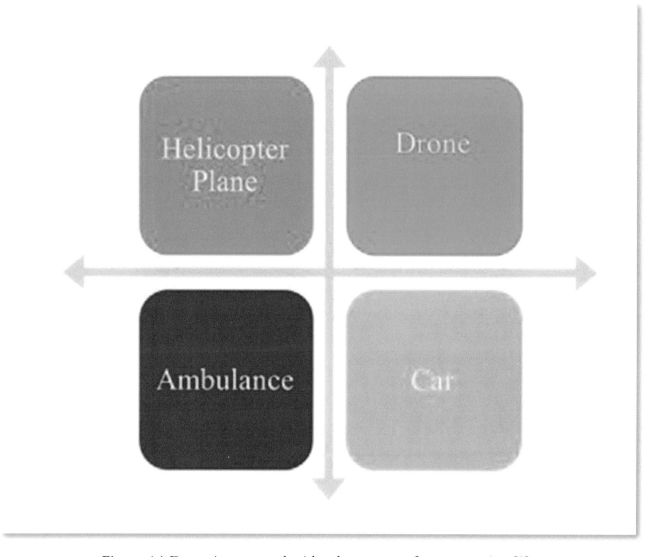

Figure 4.1 Drone is compared with other means of transportation [2].

Drones are increasingly being used as innovative tools for healthcare. Hospitals are experimenting with drones for numerous tasks that would often take twice as long for humans to do. The days are fast approaching when cars and trucks will be replaced by drones for moving things across hospital campuses.

This chapter discusses the use of drones in the field of healthcare. It begins by discussing the concept of drones. It provides some common applications of drones in healthcare. It covers the adoption of healthcare drone around the world. It discusses their benefits and challenges in healthcare. The last section concludes with comments.

4.2 CONCEPT OF DRONES

Drones are autonomous robots that fly in the sky. They may also be regarded as pilotless aircrafts that were initially used by the military, but are now used for scientific and commercial purposes. The word "drone" was coined due to the similarity of its sound to a male bee. Drones are pilotless aircraft and are formally known as either unmanned aerial vehicles (UAVs) or unmanned aircraft systems (UAS). Drones are also called "remotely piloted vehicle" or "unmanned aerial systems." Drones were first used in the 1990s by military organizations. The notion of drones began around 1918 when the US Navy commissioned Charles Kettering built a militarized unmanned aerial vehicle (UAV). Their original use was to take strategic pictures for the military. From the beginning of the 21st Century, civil activities of drones started to get more attention.

Drones are classified in different ways: size, weight, flight time, commercial or military, and cost. The US Federal Aviation Administration (FAA) defines consumer and commercial drones as those that weigh less < 1.0 lb (0.45kg) with approximately a maximum 500 m altitude and 2km range from the base operator.

A drone is a pilotless aircraft that operates through a combination of technologies, including computer vision, artificial intelligence, object avoidance tech, automation, robotics, and miniaturization.

Drones in the medical field have been used for blood delivery, food delivery, and *package delivery. They* are now used in different fields including transportation, healthcare, news media, commerce, safety and security, disaster management, rescue operations, crop monitoring, weather tracking, environmental protection, intelligence gathering, surveillance, aerial photography, express shipping, recreation, agriculture, wildlife, military, law enforcement, home, cemetery management, power, infrastructure,, telecom sectors, marine, weather forecasting, sports, space, insurance, hotels, journalism/news coverage, and logistics [3,4]. Drones have been used by the military in combat and for humanitarian aid. Drones have emerged as interaction devices in home and research applications. A drone can be used as a companion, personal drone, agent, sensing tool, delivery tool, ambulance drone, etc. Drones are commonly used by hobbyists just for the fun of it. Healthcare is benefitting from this disruptive technology. Thus, we have different kinds of drones: military or armed drones, healthcare drones, medical drones, biomedical drones, smart drones, humanitarian drones, collaborative drone, ambulance drones, courier drones, nano or micro or mini drones, etc.

The technology involved in drone construction is impressive. Modern drones are empowered by sensor technology. Accelerometers are often used to determine position and orientation of the drone in flight. Inertial measurement units combined with global positioning systems (GPS) are critical for maintaining direction and flight paths [5]. Drones have to provide a reliable connection regarding tracking as well as remote control purposes and communicate this with users. They can maneuver unobtrusively above the ground towards a target without disturbing human movement on the ground.

Examples of drones used in healthcare include [6]:

- Seattle's Village Reach
- Flirtey
- Ehang
- ZipLine
- Tu Delft
- Google Drones
- Project Wing

- HiRO (Healthcare Integrated Rescue Operations)
- Vayu Drones

Are these the name of health care companies? It's not clear what this list really is

4.3 APLICATIONS

Some drone applications involve surveillance using an on board camera. Drones can also deliver small loads. They can gather real time data effectively. Drones have been used in different areas in healthcare including transfer of blood products, enhance search and rescue efforts, collection of different types of data, delivery of rural healthcare, offer remote telemedicine or patient care at home, transport samples and deliver blood, vaccines, medicines, organs, life-saving medical supplies (ie, automated external defibrillators), and equipment. Healthcare drones are regarded as "medicine from the sky." Figure 4.2 shows some medical uses of drones [7].

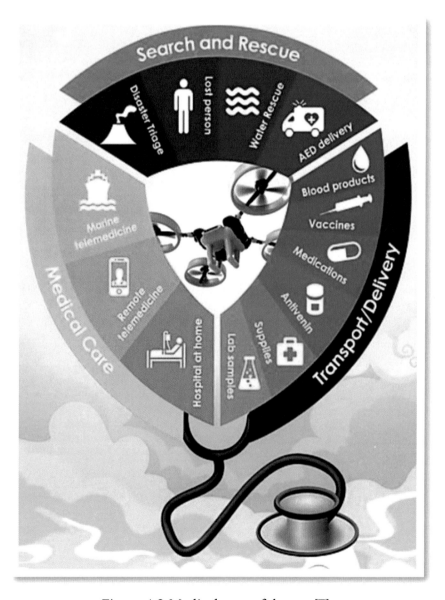

Figure 4.2 Medical uses of drones [7].

Some other potential uses of drones in healthcare include the following [8,9]:

- *Rural Healthcare:* This is perhaps the most popular use of healthcare drones. Patients often die due to lack of access to a basic medical product. Drones can be used to deliver blood, vaccines, birth control, and other medical supplies to rural areas or areas with mountains, deserts, or forests, roads are impassable and take long-distance travel. This could mean the difference between life and death. Health Wagon, a clinic based in Southwest Virginia, partnered with NASA researchers to fly the first drone approved by the Federal Aviation Administration (FAA) to deliver medication to rural Americans, particularly Native American populations. Zipline started transporting blood and vaccines to remote areas in Rwanda in harsh weather.
- *Care Delivery*: Drones can be used to transport blood, medical supplies, and water. This is a big step forward in saving time and lives. Drones can deliver supplies and medications from floor to floor or building to building. Drones can make it easier for clinicians to administer care to these patients in their homes by sending them necessary medications and other essentials for treatment. They are enabling healthcare delivery by providing faster response times, reduced transportation costs, and improved medical services to remote areas. Figure 4.3 shows a healthcare drone for delivery and pickup service in rural area [10].

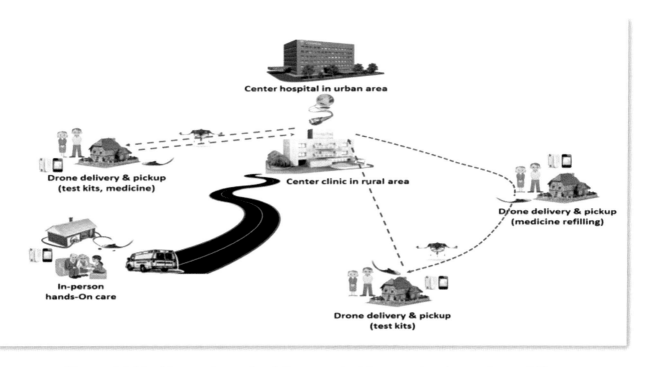

Figure 4.3 Healthcare drone for delivery and pickup services in rural area [10].

- *Bedside Medication Delivery:* Instead of nurses bringing medication to patients, drones could do the same job. Nurses and pharmacists can work more efficiently as supplies can be summoned to the bedside of a patient being cared for in the hospital or home.
- *Healthcare Transportation:* Drones are used as a means of transport. Drones can be used to transport urine, blood, sputum, and other samples from patients' rooms to hospital labs for testing. An example is the transportation of organs between two hospitals in the same city. Figure 4.4 illustrates typical medical drone transporting human organ [11].

Figure 4.4 A typical medical drone transporting human organ [11].

- *Helping the Elderly:* Research is being done to use drones to benefit the growing senior population. Drones can be used to create ambient assisted living environments for the elderly. This is an efficient and cost-effective way of providing care for the elderly while extending their independent life. Drones can be used to deliver medication refills and pick up blood and urine tests [12].
- *Emergency:* Time is critical in emergency situations, especially when patients need blood, intravenous fluids or medications. For example, time is crucial in out-of-hospital cardiac arrest. Emergency medical services can sometimes have a long response time. Drones have been used to deliver supplies during an emergent lifesaving situation. A drone can serve as an ambulance during emergencies or be a viable alternative to ambulance response (how specifically??)[13]. The drones always outperform the ambulances. Two hospitals in Switzerland started using drones to deliver blood samples.

Drones are used in other areas of healthcare including epidemiology, telemedicine, transfer units, humanitarian logistics, disaster relief, delivering aid packages, adverse weather conditions, and transporting medical goods over urban traffic jams.

4.4 GLOBAL HEALTHCARE DRONES

Drones are a rapidly developing technology with increasing worldwide applications.

Healthcare drones are increasingly being used around the globe, especially in the following places:

- *Africa:* In spite of its poverty and poor infrastructure, subSaharan Africa has led the way in the adoption of mobile banking and healthcare, and now medical drones [14]. Necessity is the mother of invention; the rapid adoption of drone technology in sub-Saharan Africa is exemplary. The drone technology caught on so quickly in developing nations such as Rwanda, Tanzania, and Ghana with poor roads and lack of accessibility. These are among a group of early adopters of transport drones for healthcare systems in sub-Saharan Africa, with the first drone airport established in Rwanda. Figure 14.5 shows World Bank president, Jim Yong Kim, hailing Rwanda's use of drones in healthcare delivery [15]. The drone is operated by Zipline, a company based in San Francisco, California, focused on lifesaving medical supplies. Zipline argues that minimizing waste in the medical system will help the drones pay for themselves. Zipline plans to expand its operations to more African nations. The United Nations has used drones to drop condoms over rural parts of Ghana, where a fraction of women have access to contraceptives to prevent the transmission of sexual transmitted diseases. Threats of criminality, security breaches, and loss of privacy make drones too risky in developed countries. The World Economic Forum in partnership with Zipline and the World Bank has been raising awareness for using drones in Africa and beyond [16]. Tanzania is planning to have the largest drone healthcare delivery operation in the world. In spite of the fast adoption of drones in the continent, drones are manufactured elsewhere, e.g. USA and China.

Figure 14.5 World Bank president hails Rwanda's use of drones in healthcare delivery [15].

- *United States*: Commercial use of drone in the US has been held back by regulations. The use of drones at hospitals in the US is relatively new. Some are of the opinion that deploying drones for public safety and humanitarian aid will be key to public acceptance in the United States. FAA defines consumer and commercial UAS as those that weigh less than 1.0 lb (0.45kg) with approximately a maximum 500 m altitude and 2km range from the base operator. FAA approved open drone fields (to indulge drone hobbyist) are dotted around the United States. The US Department of Defense is investigating drones as a battle space platform for medical logistics delivery. The National Aeronautics and Space Administration (NASA) recently tested a medical supply delivery (included medications for asthma, hypertension and diabetes) to a small clinic in rural Virginia using a drone.
- *India:* In India, there is a shortage of safe blood in hard-to-reach places. In rural India, the delay in diagnosing diseases due to lack of laboratory facilities nearby can be overcome by the use of drones. In December 2018, India's Ministry of Civil Aviation released a comprehensive framework for the operation of drones. In 2020, the government of Telangana (a state in India) in collaboration with the World Economic Forum has formalized the plan called "medicines from the sky." Telangana is known for using technology to improve the lives of the citizens. The National Disaster Management Authority (NDMA) has started using drones to handle disaster relief and rescue in India. Drones could be a gamechanger in states that are particularly geographically difficult to traverse.
- *China:* Consumer delivery was challenging in some areas in China. Drones sped the delivery of much-needed consumer goods replacing hours-long drives. The outbreak of coronavirus in China has led to significant adoption of drones. China is known for building the most small drones.
- *Pakistan:* Consumers in Pakistan perceive privacy issues as a primary concern in relation to drone delivery, and unfortunately the Pakistani government does respect the rights of its citizens to privacy [17].. The typical Pakistani consumer now increasingly prefers variety, convenience, and ease. Drones can be used by businesses for making home delivery of packages in urban areas. They are also used for transporting medical supplies to remote areas that are hard to reach by road.
- *Singapore:* Tiny high-tech Singapore, with a population of 5.6 million, is ultra-modern, well-ordered and a tightly regulated nation. Drones are being used across the nation to deliver life-saving medical supplies. Its uses could include transporting blood samples, delivering emergency medical supplies, responding to security incidents, and in the (military?) force's operations. The drones would be operated remotely by pilots and be able to travel relatively long distances across the city-state [18].

4.5 BENEFITS

Drones are increasingly being used as innovative tools for healthcare. They address a specific niche application and provide an interesting non-traditional solution. They present a tremendous opportunity to address supply chain shortcomings in the healthcare sector. They can overcome the logistic challenges since they are not subjected to traffic delays. They have become a solution to transport challenges for medical products such as emergency blood supplies, vaccines, medicines, diagnostic samples, and organs.

In places like Rwanda, Ghana, and the Philippines, drones are already being used to transport lifesaving medical goods such as blood, drugs, and other critical medical supplies on a regular basis. The major benefit of using drones is that they decrease the travel time for diagnosis and treatment. Other benefits include the following [19].

- *Cost Advantages:* Drones have cost benefits. Drones have gained their popularity due to their affordability. Lightweight drones are probably less expensive than a car or motorcycle and faster. A lost drone is less expensive than a lost helicopter or airplane. Almost any hospital can use this modern healthcare technology to its advantage. The operating costs of drones can be reduced by working with drone battery manufacturers to create rechargeable batteries.
- *Better Mode of Transportation*: Compared to ground transportation, the benefits of drones include avoiding traffic in populous areas and circumventing dangerous fly zones in war-torn countries. Drones could also mitigate safety issues that arise with traditional modes of transport. Traffic conditions (such as icy roads, foggy skies, rush hour, mountains, canyons, or snow-covered ground) impose dangerous slowdowns in delivery. Drones could offer effective alternatives in those hazardous situations. Inaccessible roads no longer have to prevent healthcare delivery.
- *Can Save Lives:* Drones can also be used as lifesavers. For example, mountain rescuers use drones in hard to reach areas. They can also make it possible to identify victims in delicate situations.
- *Saves Time:* With just a push of a button, users can fly a drone and start recording moving images to its surroundings in a lesser time. Users will no longer have to wait for a photographer or digital artist to record images. Drone photography is a dynamic and fun new type of photography.
- *Easily Deployable:* Drones can be deployed and operated with relatively minimal experience. They are becoming accessible to a wide range of healthcare purposes. They can fly lower and in more directions, allowing them to reach traditionally hard-to-access areas.

Ground-to-drone communications must be protected to prevent hackers from using their data for evil purposes. Drones can carry defibrillators to heart attack victims faster than an ambulance. A drone that could bring medical aid to people in distress or rural areas. Drones can be useful during urgent and non-urgent medical disaster relief efforts.

Other benefits of using drone in healthcare delivery include [7]:

- Increased ability to reach victims who require immediate medical attention
- Increased ability to care for the elderly
- Increases the efficiency of providing care to patients in remote locations

4.6 CHALLENGES

The rise of drones in various fields, from using them for recreational use (ie, play toys) to mass destruction weapons (ie, military), brings its own challenges. Restrictions on drones have limited their use in healthcare:

- *Privacy and Safety Concern:* Privacy and safety are major issues with drones. Drones can invade an individual's privacy and gather sensitive data about the person. As a relatively new technology, there are no regulations regarding drone safety. Security is an important part of patient safety.

Security refers to measures used to prevent health data from unauthorized access. Many things can go wrong when an autonomous flying vehicle flies across a busy road. The widespread adoption of drone in healthcare can create security concerns for healthcare practitioners: physicians, nurses, information technology, administrators, and healthcare management.
- *Regulatory Issues:* Healthcare drones raise some regulatory issues. A major hurdle in the use of healthcare drones is the need for permission from Aviation authorities. In the United States, the Federal Aviation Administration (FAA) provides license to fly drones, provided the drone meets certain requirements. In Europe, the European Aviation Safety Agency (EASA) is the legislative body with regulatory authority over drone usage. Regulations in some nations do not allow for a drone above certain size and weight.
- *Technical Limitations*: Drones cannot carry heavy loads or deliver goods long distance like commercial planes and helicopters. Weather and electromagnetic interference (EMI) can affect the performance of drones. Drones have limited carrying potential. They are unable to provide door-to-door services, which means the technology is limited to observation benefits only. Another major drawback is that a drone is powered by batteries and can exhausted after 15 minutes of flight.
- *Hacking:* Drones make it possible for unidentifiable individuals to spy on us from a distance. The vulnerability of mobile devices in the healthcare environment makes them an attractive target for hackers [20]. Ground-to-drone communications must be protected so that hackers do not hijack the drones.

Other challenges associated with using drone in healthcare delivery include [7]:

- Maintaining the integrity of specimens during delivery
- The need for special equipment (packaging)
- Payload capacity
- Limited battery life and weight carrying constraints
- Security for controlled substances
- Regulations on a local, state and federal level
- Consumer demand

These challenges have caused healthcare applications of drone technology to be slower to develop.

4.7 CONCLUSION

The age of drones has arrived and drones are here to stay. Autonomous flying is being applied in several industries ranging from transportation to law enforcement and defense.

They are emerging as a new healthcare tool that can help mitigate logistical problems and make healthcare more accessible and save lives. They are increasingly being used for healthcare purposes around the world. They now let us do things we could not do before.

Drones could ultimately be a healthcare technology game-changer. In the not-too-distant future, drones could become an essential part of the hospital delivery healthcare services within the hospital and to your home. The future of drone deliveries of all kinds in America and elsewhere rests with regulators. More information about the use of drones can be found in the book in [21] and the following related journals:

- *Journal of Unmanned Vehicle Systems*
- *Journal of Intelligent Robot System*

REFERENCES

[1] P. Kardasz et al., "Drones and possibilities of their using," *Journal of Civil & Environmental Engineering,* vol. 6, no.3, 2016.

[2] "Drones and blood transportation: Will drone impact society?" https://mbamci.com/drones-and-blood-transportation/

[3] S. H. Alsamh et al., "Survey on collaborative smart drones and Internet of things for improving smartness of smart cities," *IEEE Access,* vol. 7, 2019, pp. 128125- 128152.

[4] "38 Ways drones will impact society: From fighting war to forecasting weather, UAVs change everything," January 2020, https://www.cbinsights.com/research/drone-impact-society-uav/

[5] C. Winkler, "Sensor solutions play critical roles in enabling innovation in drone," June 2016, https://www.designworldonline.com/sensor-solutions-play-critical-roles-in-enabling-innovation-in-drones/

[6] F. Scott, "Drones in healthcare: The rise of the machines," http://csohio.himsschapter.org/sites/himsschapter/files/ChapterContent/csohio/Drones%20in%20Healtcare%20Rise%20of%20the%20Machines.pdf

[7] "A role for drones in healthcare," https://www.dronesinhealthcare.com/

[8] J. White, "How drones could revolutionize care delivery at your hospital," June 2019, http://www.healthcarebusinesstech.com/drones-care-hospital/

[9] M. Blau, "Condom drops and airborne meds: 6 ways drones could change health care," June 2017, https://www.statnews.com/2017/06/13/drones-health-care/

[10] S. J. Kim et al., "Drone-aided healthcare services for patients with chronic diseases in rural areas," *Journal of Intelligent Robotic System,* (2017) vol. 88, 2017, pp. 163–180.

[11] "Drone delivery models for healthcare," https://adalidda.com/posts/9ckvrmyhrA7Rp7zHK/drone-delivery-models-for-healthcare

[12] R. Sokullu, A. Balcı, and E. Demir, "The role of drones in ambient assisted living systems for the elderly," in I. Ganchev et al. (eds.), *Enhanced Living Environments: Algorithms, Architectures, Platforms, and Systems*. Springer, 2019, pp 295-321.

[13] P. V. de Voorde, "The drone ambulance [A-UAS]: Golden bullet or just a blank?" Resuscitation, vol. 116, July 2017, pp. 46-48.

[14] B. McCall, "Sub-Saharan Africa leads the way in medical drones," *World Report*, vol 393, January 2019, pp. 17-18.

[15] "World Bank president hails Rwanda's use of drones in healthcare delivery," http://www.xinhuanet.com//english/2017-03/22/c_136148796.htm

[16] R. Sengupta, "Drones deliver medicines in Africa," June 2019, https://www.downtoearth.org.in/news/health-in-africa/drones-deliver-medicines-in-africa-64832

[17] "Singapore to use drones to transport medicine," July 2018, http://www.healthcareasia.org/2018/singapore-to-use-drones-to-transport-medicine/

[18] R. Khan, S. Tausif, and A. J. Malik, "Consumer acceptance of delivery drones in urban areas," *International Journal of Consumer Studies*, vol. 43, no. 1 January 2019, pp. 87-101.

[19] Mario, "The pros and cons of drones (UAVs)," https://www.dronetechplanet.com/the-pros-and-cons-of-drones/

[20] A. Alexandrou, "A security risk perception model for the adoption of mobile devices in the healthcare industry," *Doctoral Dissertation*, Pace University, July 2015.

[21] T. Kille, P. R. Bates, and S. Y. Lee, *Unmanned Aerial Vehicles in Civilian Logistics and Supply Chain Management*. IGI Global. 2019.

CHAPTER 5

DRONES IN EDUCATION

"Education is our passport to the future, for tomorrow belongs to the people who prepare for it today."

– Malcolm X

5.1 INTRODUCTION

Drones are used as tools in various industries such as shipping, agriculture, construction, entertainment, and more. Drones have incredibly wide-ranging uses. Industry professionals are using drones to take artistic images or to gather valuable data. The use of drones has increased in the science, technology, engineering, and mathematics (STEM) professions. The sooner our children can engage with the technology, the better equipped they will be to innovate with it into the future. Incorporating drone information and exercises into the curriculum for STEM students is important for career preparation because of the widespread integration of drones as tools across many fields. For teachers, using drones in the classroom open up a new set of opportunities to make classes more relevant and engaging.

Drones are naturally fun and educational. Using them can keep students engaged. However, bringing a drone into your classroom is not just about teaching students to become drone pilots. The drone is simply the tool to be able to teach essential skills of problem-solving, digital competence, coding, and creativity. Covering drone technology in the classroom helps prepare students for jobs in the fast-growing, multi-billion dollar industry. There are lessons that can be taught better with the use of drones. Educators must look for those teachable moments and challenge the problem-solving skills and creativity of students. With a drone, the sky is the limit for what students can learn.

While originally used for military purposes, drones now boast diverse capabilities and usages. They can be very powerful interactive learning tools for students at any level. Drones can help teach a wide range of concepts and lessons that otherwise might be difficult for students to understand. Most drones are easy to learn to fly, and many are inexpensive, making them accessible to everyone. Teaching with drones holds many possibilities, from introducing kids to piloting or coding basics to exploring drone uses and careers. Drones are excellent complementary tools for STEM education. From coding to critical thinking, drones are perhaps the best option schools have at their disposal for break-through STEM learning [1].

This chapter explores the various applications of drones in education. It begins with describing what a drone is. It presents some applications of drones in education, with emphasis on their use in high school. It highlights some benefits and challenges of drones in education. The last section concludes with comments.

5.2 WHAT IS A DRONE?

The FAA defines drones, also known as unmanned aerial vehicles (UAVs), as any aircraft system without a flight crew onboard. Drones include flying, floating, and other devices, including unmanned aerial vehicles (UAVs), that can fly independently along set routes using an onboard computer or follow commands transmitted remotely by a pilot on the ground. A typical drone is shown in Figure 5.1 [2]. Drones can range in size from large military drones to smaller drones. Drones, previously used for military purposes, have started to be used for civilian purposes since the 2000s. Since then, drones have continued to be used in intelligence, aerial surveillance, search and rescue, reconnaissance, and offensive missions as part of the military Internet of things (IoT). Today, drones are used for different purposes such as aerial photography, surveillance, agriculture, entertainment, healthcare, transportation, law enforcement, etc.

Figure 5.1 A typical drone [2].

Drones work much like other modes of air transportation, such as helicopters and airplanes. When the engine is turned on, it starts up, and the propellers rotate to enable flight. The motors spin the propellers and the propellers push against the air molecules downward, which pulls the drone upwards. Once the drone is flying, it is able to move forward, back, left, and right by spinning each of the propellers at a different speed. Then, the pilot uses the remote control to direct its flight from the ground [3].

Drone laws exist to ensure a high level of safety in the skies, especially near sensitive areas like airports. They also aim to address privacy concerns that arise when camera drones fly in residential areas. These include the requirement to keep your drone within sight at all times when airborne. In the United States, drones weighing less than 250g are exempt from registration with civil aviation authorities. If your drone exceeds 250g in weight, you will also require a Flyer ID, which requires passing a test [4]. It is necessary to register as an operator, be trained as a pilot, and have civil liability insurance, in addition to complying with various flight regulations, and those of the places where their use is permitted.

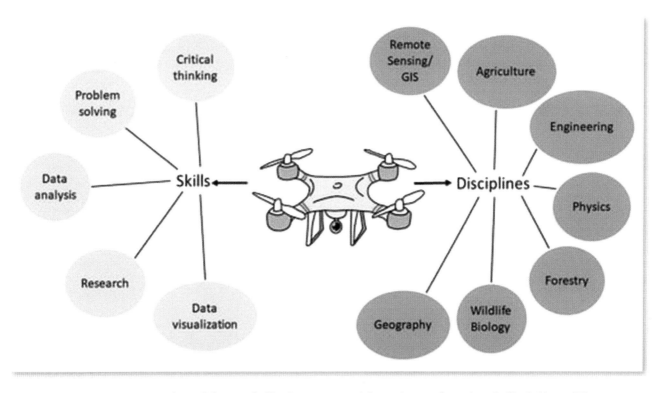

Figure 5.2 Examples of drone skills that can enrich various educational disciplines [5].

5.3 APPLICATIONS OF DRONES IN EDUCATION

There are several ways drones are being utilized by teachers across the world. Figure 5.2 shows examples of drone skills that can enrich various educational disciplines [5]. The following applications are typical [6,7]:

- *Media:* Today's students love digital media. Drones have emerged as powerful tools for students looking to explore the world of digital media and photo/video editing. Integrating drones into educational curricula allows students to learn to navigate and control the drone, understand the principles of framing and composition, and discover the unique perspectives that aerial photograph. Moreover, drones provide a real-world platform for students to experiment with various camera settings, lighting conditions, and angles, which are fundamental skills in digital media production. Ultimately, incorporating drones into the learning process empowers students to transform their visions into captivating visual narratives.

- *Environmental Science:* This is another subject today's students find fascinating and one with which drones can be utilized to further learning. Drones have revolutionized environmental education by providing students with an unparalleled perspective of the natural world. Through drone technology, students can witness ecosystems, landscapes, and wildlife from a bird's-eye view.

- *Photogrammetry:* Drones have emerged as invaluable tools for introducing students to the fascinating world of photogrammetry. This is a technique that involves capturing precise measurements and creating detailed 3D models from aerial imagery. By piloting drones equipped with high-resolution cameras, students can collect a series of overlapping images of a specific area or object. Moreover, they gain an understanding of how this technology is applied in various fields, from urban planning and archaeology to forestry and engineering. This experience empowers them to explore new dimensions in spatial data and opens doors to exciting career opportunities in fields that rely on accurate geospatial information.

- *Soft Skills:* Beyond the academic benefits above, drones can also be a key tool in helping children develop essential social and emotional skills. While planning and executing flights, students also utilize interpersonal soft skills like collaboration, negotiation, and active listening. The ability to work with others, clearly communicate, and remain focused will always come into play. Through team-based projects, students gain abilities such as collaboration, flexibility, self-learning, and empathy.

- *Coding:* Coding is one of the most popular jobs in this modern age. Learning how to code can be less intimidating if students use drones for hands-on learning experience. Many educators are using drones to engage students into writing programs that allow drones to conduct autonomous flights. Some classrooms are utilizing the popular coding language Python, creating code to pilot their drone around a gymnasium. Some teachers use drones in programming courses, using drones to engage students into writing programs that allow drones to conduct autonomous flights.

- *Physical Activities:* Most children between the ages of 11 and 17 are not as physically active as they should be. The number of children with obesity in the USA has tripled since the 1970s. The desire for sports and spending time outside has been replaced by video games, mobile phones and interactive tablet devices. Teachers can use drones to get the students to base physical activity around interaction with the drone.

- *Teaching Math:* Math has been a challenging subject for most students. Using learning tools such as drones can give a real world application to mathematical problems and equations, helping students realize the great power of this subject. Schools around the world are experimenting with drones to instill basic trigonometry concepts by following the path of different drones.

- *Teaching Geography:* Drones can be programmed to fly autonomously to capture a specific region. They are increasingly being used as a cost-effective way to gather geospatial data. Drones can be used to capture footage and videos of locations students might not normally be able to access. This kind of data relates to creating and reading topographical maps.

- *Teach Science:* There are many creative ways you can use drones to teach science. You can use drones to help them understand cell structure by making a large scale model and then zoom in and out to show individual parts and the larger structure. Laws of physics can also be taught with the help of drones. Options are numerous, and can bring a lot of fun into science classes.
- *Social Learning:* Drones give students glimpses of themselves and their place in the world. This technology could help students visualize themselves as being a part of something greater.

Figure 5.3 How drones are used in the classroom [7].

5.4 APPLICATIONS IN SCHOOLS

Drones are widely used as great tools in education. They are easy to integrate into every aspect of education and real life. A typical example of how drones are used in the classroom is shown in Figure 5.3 [7]. By incorporating drones into the learning process, students not only develop technical proficiency but also cultivate a deeper appreciation for the transformative potential of photogrammetry in research, design, and analysis. Specific applications of drones in schools and colleges include [6]:

- *K-12 Schools:* K-12 schools across the world are beginning to integrate drones into their curriculums. Drones provide something tangible and interesting, instead of boring students with abstract, hard-to-get textbook examples. As typically shown in Figure 5.4, students learn to operate drones safely in manual and autonomous flight scenarios [8]. Considerations for the drones suitable for elementary schools include [9]:

Figure 5.4 Students learn to operate drones safely [8].

- ➢ Drones with cameras for kids
- ➢ Rechargeable batteries
- ➢ Flight mission apps for sub-2kg drones
- ➢ Educational coding apps for microdrones
- ➢ Time it takes to learn to fly
- ➢ Cost of a drone
- ➢ Drone app integration
- ➢ Level of skill required
- ➢ Pre and post-sale support
- ➢ Ratio of drones to students

These features and considerations should help the educator decide what drone to buy and use.

- *Higher Education*: University courses can greatly benefit from the introduction of drones. Students from fields as varied as journalism, engineering, botany, humanities, and forestry have started using this technology to enhance their professional training and to add real-world examples to their studies. Drones can be used to teach skills valued in the humanities. For social studies classes, drones can be used to study local geography, cartography, and the history of their communities. The more advanced the course, the more possibilities for drones to take center stage, with the ability to use modern technology to convey complicated physics and calculus concepts. Some courses are used to prepare students for careers in UAS or drones.

The most common drones used in education settings fall into two categories: microdrones (under 250 grams) and sub-2 kg drones. A microdrone is usually around $150-$300, whereas a decent sub 2kg drone is currently from $600 to over $3,000. It is recommended to fly with the smallest drone possible to achieve the chosen learning outcomes. This reduces the risk profile and the cost of purchase.

5.5 BENEFITS

Drones can be used to help enhance orientation skills, motor skills, and even give students a better understanding of how the world around us works. By far, the greatest advantage of integrating drones in the classroom is the fact that they are incredibly fun and educational. Drones foster intellectual curiosity and creativity. They still maintain the novelty of being new and unfamiliar technology for most people, so they tend to draw attention. They are inexpensive and accessible, making them a good fit for educational settings. They can also be used to monitor student performance during physical activities. Figure 5.5 shows ten ways to use drones in education [10].

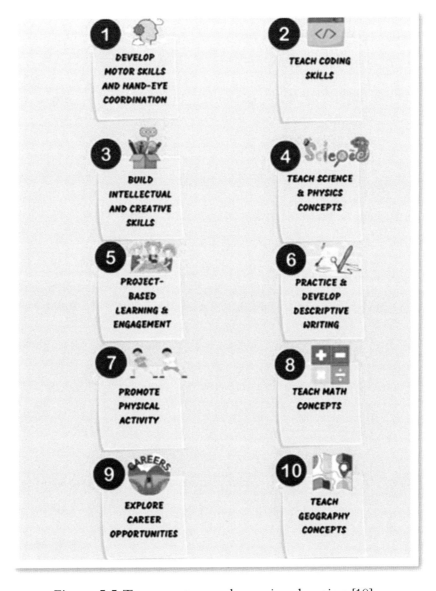

Figure 5.5 Ten ways to use drones in education [10].

The economic benefits of drone technology include [9]:

- Creation and support of 5,500 full time equivalent jobs
- $14.5 billion increase in GDP over the next 20 years – of which $4.4 billion would be in regional areas across New South Wales, Queensland and Victoria
- Cost savings of $9.3 billion over the next 20 years, with $2.95 billion of this in the agriculture, forestry and fisheries industries, $2.4 billion in mining and $1.34 billion in construction
- Bringing on the next wave of "jobs you haven't thought of" as our primary school students of today enter the workforce of 2030.

Drones can be used in education to help students develop a variety of skills, including:

- *Coding:* Drones can be used as a hands-on way to learn coding, as students can see the results of their coding in the drone. Students can see the results of their coding in real time, which can help them learn computational thinking.
- *Critical Thinking:* Drones can help students visualize problems from different perspectives, which can lead to more creative solutions.
- *Project Management:* Drones can help students learn project management skills.
- *Technology Proficiency*: Drones can help students learn about technology.
- *Collaboration*: Flying drones can encourage teamwork and social interaction.
- *Confidence:* Flying drones can help students build confidence.
- *Writing:* Drone imagery can be used as the basis for descriptive writing exercises.

The benefits of drone technology, particularly as a part of a STEM curriculum, are well worth the investment while being crucial in preparing students for the future. Drones can be used in education to teach a variety of subjects and skills, including STEM, coding, and critical thinking:

- *STEM Learning*: Drones can be used to teach STEM concepts like aerodynamics, aviation, programming, and data analysis. The data collected from drones can help students study how a phenomenon relates to its environment, people, and time. Students can explore coding, aerodynamics, data analysis, and more, thereby enhancing their understanding and interest in STEM fields.
- *Mathematical Skills:* Drones can help students learn math concepts in a more engaging way, and can help them develop critical thinking and problem-solving skills.
- *Physical Activities*: Flying drones can encourage social interaction and teamwork, and can be used to monitor student performance during physical activities.
- *Physics:* Drones can be used to illustrate how physical laws work, such as the law of gravity.
- *Problem-solving:* Drones can help students visualize problems from different perspectives, which can help them think critically and creatively.
- *Hands-on Projects*: Hands-on training courses can provide a safe environment for beginner pilots to practice flying drones. The hands-on nature of drones makes abstract concepts more accessible and stimulates curiosity and innovation.

5.6 CHALLENGES

In spite of the high perceived need to incorporate drone into education, several barriers have prevented widespread drone training. A lot of schools cannot incorporate drones into their curriculum due to a lack of subject familiarity or access to technology because of funding constraints. Other challenges to using drones in education include the following [5,11]:

- *Safety:* Drones can be dangerous in populated areas, such as college campuses. Pilots need to be aware of the risks to people, buildings, and equipment. Schools may want to get liability insurance to protect against third-party drone operation.
- *Regulations:* Educators may not be fully aware of the regulations around drones, including privacy, safety, and airspace considerations. Regulations for drones restrict who can legally fly a drone and the flight location. The regulations and licensing requirements could impede the incorporation of hands-on drone exercises for many educators.
- *Integration Challenges*: It can be difficult to fully integrate drones into the educational process. Some studies have identified limitations such as small sample sizes, limited exposure time, and lack of population diversity.
- *Class Size:* Equipment, maintenance, safety, and technical requirements can limit the number of students in a drone activity. For example, some say that lower secondary classes should have 11–20 students, and upper secondary classes should have 21–30 students.
- *Curriculum:* Educators are working to develop curricula that meet the needs of a variety of industries. It can also be challenging to align drone courses with other programs at a university.
- *Teacher Skills*: Teachers need to acquire new skills to handle the challenges of integrating drones into the educational process.
- *Drone Selection*: It is important to choose the right drone for the task. For example, students should start with a smaller, low-risk drone, such as a microdrone.
- *Support:* It is important to have local support for tech and warranty issues. Drones can be a powerful tool for STEM education. They can help students develop critical thinking and problem-solving skills, and deepen their understanding of STEM principles
- *Awareness:* There is still not broad awareness of the commercial applications of drones.
- *Learning Outcomes:* There is a gap in evaluating the learning outcomes of drone education practices.
- *Battery Management:* Drones require effective battery management. Some drones require charging the battery while it is inside the drone, which means you cannot fly while it is charging.
- *Environmental Impact:* Drones can crash or get stranded in remote areas, which can damage the environment. Drone batteries can also be harmful if not disposed of properly.

5.7 CONCLUSION

It has become increasingly common in recent years for teachers to utilize drones as learning tool within the classroom. This provides students with the opportunity to learn about technology that they may wish to base their future career on. Drone are interactive, instantly engaging, and offer opportunities for multidisciplinary lessons.

All drone operations must be completed in line with local laws and regulations and within a safe and responsible manner. Drones flown for educational and research purposes are regulated under the Exception for Limited Recreational Operations of Unmanned Aircraft. They may be operated under § 44809 for qualifying educational organizations. Educational institutions, including primary and secondary educational institutions, trade schools, colleges, and universities are eligible to request the establishment of a FAA-Recognized Identification Area (FRIA).

Drone technology is here to stay and will continue to be part of our modern world. In no distant future, educational drones will reach a level of commonality in our lives, especially drones as an education tool in schools. For more information about drones in education, the reader should consult the books in [12-16].

REFERENCES

[1] M. N. O. Sadiku, M. Oteniya, J. O. Sadiku, and S. Abunene, "Drones in education," *International Journal of Latest Engineering Research and Applications,* vol. 9, no. 8, August 2024, pp. 1-7.

[2] "Best drones for education: from building and flying to coding," https://www.eduporium.com/blog/best-drones-for-education-building-flying-and-coding/

[3] "How drones work and how to fly them," May 2024, https://dronelaunchacademy.com/resources/how-do-drones-work/

[4] "What are the main applications of drones?" June 2024, https://www.jouav.com/blog/applications-of-drones.html

[5] M. M. Bolick, E. A. Mikhailova, and C. J. Post, "Teaching innovation in STEM education using an unmanned aerial vehicle (UAV)," *Education Sciences*, vol. 12, no. 3. 2022.

[6] "How drones can improve K-12 education." https://www.dronegenuity.com/how-drones-benefit-k-12-education/

[7] L. Wakefield, "10 Ways drones are being used in the classroom," October 2023, https://www.coverdrone.com/10-ways-drones-are-being-used-in-the-classroom/

[8] "Drones STEM education," https://nextwavestem.com/drones-stem-curriculum

[9] K. Joyce, "Drones in education: The ultimate guide for your school," March 2024, https://shemaps.com/blog/drones-in-education/

[10] A. Verweij, "Drones in K-12 education," https://edgeucating.com/drones-in-k-12-education/

[11] C. Lauter, "Confronting the challenges of drones in higher education," September 2024, https://www.commercialuavnews.com/university-roundtable-2024#:~:text=One%20of%20the%20key%20areas,awareness%20of%20their%20commercial%20applications.

[12] C. Carnahan, L. B. Zieger, and K. Crowley, *Drones in Education: Let Your Students' Imagination Soar.* International Society for Technology in Education, 2016.

[13] G. D. Hoople, A. Choi-Fitzpatrick, *Drones for Good: How to Bring Sociotechnical Thinking Into the Classroom*. Morgan & Claypool Publishers, 2020.

[14] A. Loureiro, H. R. Gerber, and M. J. Loureiro (eds.), *Handbook of Research on Global Education and the Impact of Institutional Policies on Educational Technologies*. IGI Global, 2021.

[15] H. Monthie, *Build, Code, Fly, Certify: Drones in STEM Education: Taking STEM Education to New Heights*. Independently Published, 2024.

[16] K. Yaun, *Drones in the Classroom (Inside the World of Drones)*. Rosen Young Adult, 2016.

CHAPTER 6

DRONES IN MANUFACTURING

"Manufacturing is more than just putting parts together. It's coming up with ideas, testing principles and perfecting the engineering, as well as final assembly."

—James Dyson

6.1 INTRODUCTION

Drones, once known as outdoor toys for kids and teenagers, turned out to have immense potential to transform various industries. This is due to their ability to reach difficult or dangerous areas for humans or traditional machinery to access. They have made impressive strides in various industries, with applications ranging from aerial photography to environmental monitoring. Drones embedded themselves as a crucial technology in Industry 4.0, offering a range of benefits for industries such as manufacturing, logistics, and agriculture. They have quickly become indispensable tools in the manufacturing industry and are revolutionizing the way manufacturing facilities operate. Their integration into the manufacturing industry offers immense potential for improving efficiency, safety, and cost savings.

Drones generally are unmanned aerial vehicles, with nobody on board, and the pilot is controlling the craft from the ground. They are far from being just fancy flying cameras. They have been used for all kinds of purposes. Their sophisticated technology allows them to obtain and record information where humans cannot, such as in dangerous environments and difficult to access areas. The prohibition on beyond visual line of sight operations has also been lifted for drones that weigh more than five kilograms. This allows bigger drones, ideal for commercial and industrial applications, to be used more consistently in manufacturing. Drones can also aid manufacturers in increasing compliance by recording temperature checks, production line observations and faults from the drone images.

Drones, also known as unmanned aerial vehicles (UAVs), are rapidly taking the world by storm, and their impact on manufacturing is significant. They are generally unmanned aerial vehicles. They can navigate treacherous terrain, gather critical data, and perform assessments without risking human lives. In the world of rapidly evolving technology, drone manufacturing stands out as a field combining innovation with precision. From agriculture to construction, logistics to environmental conservation, drones have found their way into diverse sectors [1].

This chapter delves into the applications of drones in the manufacturing sector. It begins with describing what a drone is. It presents some applications of drones in manufacturing. It highlights some benefits and challenges of drones in manufacturing. The last section concludes with comments.

6.2 WHAT IS A DRONE?

The FAA defines drones, also known as unmanned aerial vehicles (UAVs), as any aircraft system without a flight crew onboard. Drones include flying, floating, and other devices, including unmanned aerial vehicles (UAVs), that can fly independently along set routes using an onboard computer or follow commands transmitted remotely by a pilot on the ground. A typical drone is shown in Figure 6.1 [2]. Drones can range in size from large military drones to smaller drones. Drones, previously used for military purposes, have started to be used for civilian purposes since the 2000s. Since then, drones have continued to be used in intelligence, aerial surveillance, search and rescue, reconnaissance, and offensive missions as part of the military Internet of things (IoT). Today, drones are used for different purposes such as aerial photography, surveillance, agriculture, entertainment, healthcare, transportation, law enforcement, etc.

Figure 6.1 A typical drone [2].

Drones work much like other modes of air transportation, such as helicopters and airplanes. When the engine is turned on, it starts up, and the propellers rotate to enable flight. The motors spin the propellers and the propellers push against the air molecules downward, which pulls the drone upwards. Once the drone is flying, it is able to move forward, back, left, and right by spinning each of the propellers at a different speed. Then, the pilot uses the remote control to direct its flight from the ground [3].

Drone laws exist to ensure a high level of safety in the skies, especially near sensitive areas like airports. They also aim to address privacy concerns that arise when camera drones fly in residential areas. These include the requirement to keep your drone within sight at all times when airborne. In the United States, drones weighing less than 250g are exempt from registration with civil aviation authorities. If your drone exceeds 250g in weight, you will also require a Flyer ID, which requires passing a test [4]. It is necessary to register as an operator, be trained as a pilot, and have civil liability insurance, in addition to complying with various flight regulations, and those of the places where their use is permitted.

Most drones have a limited payload, usually under 11 pounds. Drones are classified according to their size. Here are the different drone types:

- Nano Drone: 80-100 mm
- Micro Drone: 100-150 mm
- Small Drone: 150-250 mm
- Medium Drone: 250-400 mm
- Large Drone: 400+ mm

One of the emerging trends in drone use for factories is the utilization of LiDAR technology. LiDAR stands for Light Detection and Ranging. This technology provides accurate depth information essential for understanding the three-dimensional structure of the environment. LiDAR sensors emit laser beams to measure distances to objects, creating high-resolution 3D maps of the surrounding terrain and objects. The ability to capture detailed data through LiDAR technology has opened up opportunities for better predictive maintenance, reduction in inspection times, and overall cost savings [5].

Figure 6.2 A drone in the manufacturing environment [2].

6.3 APPLICATIONS

A drone in the manufacturing environment is shown in Figure 6.2 [2]. Drones have many applications in manufacturing, including quality control, inventory management, and safety inspections. Drones are undoubtedly a game-changer for the manufacturing industry. Common applications of drones in manufacturing include the following [2,5]:

- *Factory Surveillance and Maintenance*: One of the most promising applications of drones is their use in factory surveillance and maintenance. This allows for remote inspection capabilities and offers a cost-effective solution that enhances safety while increasing productivity. Integrating drones into factory operations has the potential to improve efficiency, reduce costs, and enhance safety measures in infrastructure management. Drones can be equipped with high-resolution cameras and sensors to monitor machinery and detect potential problems before they become major. They can inspect equipment without disrupting operations, providing a more accurate assessment of the situation while ensuring worker safety. They can also inspect hard-to-reach areas, such as tall chimneys, roofs, or narrow tunnels where human intervention is difficult or dangerous.
- *Inventory Management:* Drones can potentially transform the supply chain and inventory management in manufacturing facilities. By using RFID technology, drones can perform swift and accurate inventory counts in warehouses, significantly reducing time and labor costs associated with manual stocktaking.
- *Improving Quality Control*: Manufacturers can enhance their quality control processes by employing drones to monitor production lines continuously. This proactive approach to quality control can result in fewer product returns and improved customer satisfaction.
- *Inspection and Maintenance:* Drones are equipped with cameras and sensors can capture high-resolution images and data, allowing for more detailed inspections of large and complex infrastructures such as bridges, wind turbines, and oil rigs. These inspections can be performed remotely, reducing the risk of injury to workers and increasing safety. Transporting parts between different areas of a manufacturing plant or warehouse can also be unnecessarily time-consuming. Drones can transport parts weighing up to five kilograms and fly to warehouses to collect and deliver. Figure 6.3 shows drones are used for inspection and maintenance [6].

Figure 6.3 Drone is used for inspection and maintenance [6]

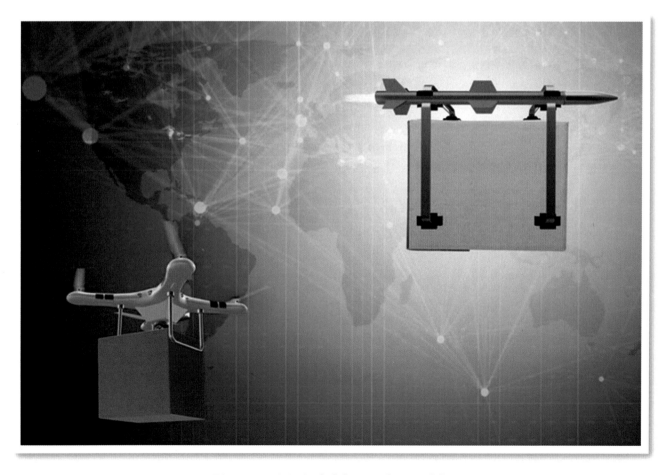

Figure 6.4 Typical delivery drones [7].

- *Automation:* Drone technology represents a revolutionary advancement in automation. There are several tasks that drones can automate in warehouses and production lines. Someone with little to no experience flying a drone can use the automation technology that accompanies the drone to carry out the flight mission. Using drones can relieve human staff from potentially dangerous jobs like retrieving products from tall racking, troubleshooting dangerous equipment during work stoppages, and responding to spills.
- *Delivery:* Drones can quickly deliver goods to customers. The journey of a product – from manufacturing, to delivery, to the customer – involves many steps. Many organizations integrate delivery drones into their assets to address this very problem. Figure 6.4 shows some typical delivery drones [7].
- *3D Printing Drones:* Also known as additive manufacturing, 3D printing is revolutionizing the drone manufacturing industry by enabling the creation of complex geometries and custom designs that were previously difficult or impossible to achieve with traditional manufacturing methods. This process involves layer-by-layer fabrication of parts directly from digital models, offering several key advantages. It allows designers and engineers to quickly produce and test prototypes. A 3D printing drone is shown in Figure 6.5 [8].

Figure 6.5 A 3D printing drone [8].

6.4 BENEFITS

Drones offer unparalleled benefits for companies in the manufacturing sector, revolutionizing the way forward in areas like surveillance, maintenance, and workflow efficiency. They can help companies save time and money, and increase compliance.

A notable advantage of using drones in industrial settings is their ability to access hard-to-reach or dangerous areas. Drones are sophisticated pieces of technology that can collect and record data humans cannot. Other benefits of using drones in manufacturing include the following [9]:

- *Efficiency:* Efficiency and cost-saving measures are at the heart of using drones in manufacturing. Integrating drones in manufacturing adds a new standard of automation and efficiency. By utilizing drones, manufacturers can reduce the risk of workplace accidents, enhance overall safety, and improve productivity. Manufacturing drones cost-effectively is essential to making the technology accessible and sustainable.
- *Safety:* The integration of drones into the manufacturing industry offers immense potential for improving efficiency, safety, and cost savings. In addition to being cost-effective and time-efficient, the use of unmanned aircraft in factories also reduces safety risks by eliminating the need for workers to climb tall structures or navigate hazardous areas.
- *Detecting Anomalies*: Through machine learning techniques, drones equipped with AI can detect anomalies that may indicate structural damage or wear-and-tear faster than a human inspector could ever do so manually. This includes monitoring changes over time that can highlight potential issues before they become major problems.
- *Accuracy:* Accuracy in drone manufacturing is paramount, as it directly impacts the performance, safety, and reliability of aerial vehicles. Precision in crafting each component ensures that the drone operates as intended, maintaining stability and efficiency in flight. This accuracy is crucial for components like propellers and motors, where even minor imbalances can lead to flight instability or mechanical failures.
- *Collaboration:* A worker doing a complex task may need support from a supervisor or team member elsewhere. A drone can help facilitate communication between these two parties by letting them see and talk to each other.
- *Compliance:* Drones can aid manufacturers in increasing compliance by recording temperature checks, production line observations, and faults from the drone images. They can also improve health and safety compliance. Plant maintenance inspections often require production to halt due to safety considerations. However, if a drone was to be used to inspect the machinery, production could continue without compromising staff safety.
- *Transportation:* Transporting parts between different areas of a manufacturing plant or warehouse can be unnecessarily time-consuming. Drones can transport parts weighing up to five kilograms and fly to warehouses to collect and deliver. The use of drones to transport parts saves manufacturers valuable time.
- *Asset Monitoring*: Drones can improve the performance of asset monitoring by using infrared and thermal technology to get accurate temperature information on machinery and production lines. For example, if temperatures were too high, drones would alert operators in time to address the issue before equipment failure could cause unwanted downtime.

6.5 CHALLENGES

In spite of the many benefits of incorporating drones into manufacturing, there are challenges that companies must address to ensure the successful implementation of this technology. Keeping pace with rapid technological changes is a significant challenge. The main problem of flying drones in a factory is their malfunctioning and crashing. Like every technical device, drones sometimes fail and fall to the ground. The greatest concern is if it falls on a worker. Other challenges include the following [5,10]:

- *Security and Privacy*: Drones can have security and privacy issues due to a lack of security and privacy considerations in their design. This includes unsecured wireless channels and insufficient computing capability. Drones can capture images and videos of sensitive information, leading to privacy concerns for employees and the company.
- *Safety:* Drones can be a hazard if not operated by trained workers. They can distract, frighten, or endanger people and birds. Drones can also threaten air traffic and workers in underground mines.
- *Cybersecurity:* As drones rely on wireless communication and data sharing, addressing cybersecurity concerns must be a top priority for manufacturers to protect their sensitive information and maintain secure operations.
- *Training and Skills Development:* Proper drone operation and maintenance training is essential for companies to maximize the benefits of incorporating this technology into their processes.
- *Regulation:* Most countries are now grappling with regulatory issues related to drones, as using drones has implications for public safety. Regulation plays a critical role in shaping the future of drone use in manufacturing since drones operate in a regulatory environment. As the use of drones in manufacturing continues to grow, regulations and policies must be implemented to ensure safety, security, privacy, and responsible use. Adhering to local and federal regulations is becoming a pivotal concern for drone manufacturers. Effective regulation will ensure the safe integration of drones into our manufacturing systems. Regulators are charged with thoroughly evaluating the implications of new drones uses, including potential safety issues, before they reach the market. In the United States, The Federal Aviation Administration (FAA) has implemented rules for commercial drone use, such as requiring pilots to obtain a remote pilot certificate and limiting altitude and speed restrictions.
- *Interoperability:* Manufacturing sites often use a diverse range of machines, equipment, and production systems, leading to issues with interoperability and the integration of new technologies. It is essential to establish a standardized ecosystem for seamless connectivity between various systems beforehand.
- *Quality Control:* Ensuring high quality and safety standards is paramount, given the potential risks associated with drone operation. For the successful implementation of AI drones, high-quality, clean, and meaningful data is vital. Since industrial settings can produce biased, outdated, or error-filled data, ensuring that the data we use to train the AI models is verified for quality is critical.
- *Supply Chain Management:* Managing the supply chain efficiently is a challenge, especially with the global nature of manufacturing. Issues like sourcing materials, managing logistics, and dealing with tariffs or trade restrictions can significantly impact production costs and timelines.
- *Cost:* Another major problem is the cost of running the drone. This is pretty much a question if you need a pilot, or if the drone runs autonomously. If you have a drone pilot, then you add labor cost to the drone when you use it and that makes it quite a bit more expensive. By using autonomous drones or incorporating advanced drone solutions into operations, companies can reduce the need for human experts and cost.

- *Weight and Strength:* Drones must be lightweight to be efficient and have good battery life, but they also need to be able to withstand environmental conditions and various stresses.
- *Design:* Industrial drone design requires a balance between the number of propellers, battery capacity, weight, and sensing and connectivity capabilities.
- *Collision Avoidance*: Ensuring that drones can reach their target without colliding with obstacles is a core problem in designing multi-unmanned aerial vehicle (UAV) systems.
- *Airspace Management:* Drones must be flown responsibly within local aviation laws and safe airspace management. This includes adhering to flight altitude limits.
- *Logistical Issues:* Drones can have difficulty maneuvering around buildings and wildlife, and delivering to specific apartments.

The challenges of drone manufacturing are significant, but with the right expertise and technology, they can be effectively addressed.

6.6 CONCLUSION

It is evident that drones will continue to play a significant role in shaping the future of manufacturing. It is never too late to start incorporating drones into your business.

The biggest obstacle, apart from certification and regulations, is that certain factories, facilities, and assembly plants may be too complex or hazardous for drones. Although there are still some limitations to commercial drone operation, including high initial costs or consistent and transparent regulations, the future of drones in manufacturing looks promising, as they are becoming increasingly popular for their ability to inspect and monitor structures and processes with greater efficiency and accuracy than traditional means. More information on drones in manufacturing can be found in the books in [11,12].

REFERENCES

[1] M. N. O. Sadiku, P. A. Adekunte, and J. O. Sadiku, "Drones in manufacturing," *International Journal of Trend in Scientific Research and Development*, vol. 8, no. 4, July-August 2024, pp. 1085-1090.

[2] "Transforming operations with drones: 10 smart drone applications in manufacturing," March 2024,
https://dac.digital/drone-applications-in-manufacturing/

[3] "How drones work and how to fly them," May 2024,
https://dronelaunchacademy.com/resources/how-do-drones-work/

[4] "What are the main applications of drones?" June 2024,
https://www.jouav.com/blog/applications-of-drones.html

[5] "Drones in manufacturing: A game-changer for industry,"
https://viper-drones.com/industries/infrastructure-drone-use/manufacturing/#:~:text=The%20integration%20of%20drones%20into,on%20manufacturing%20is%20no%20exception.

[6] "Drones' critical role in Industry 4.0,'
https://consortiq.com/uas-resources/drones-critical-role-in-industry-4-0#:~:text=Drones%20can%20quickly%20deliver%20goods,constant%20improvement%20and%20great%20efficiencies.

[7] J. Marsh, "The developing role of drones in manufacturing," August 2022. https://www.roboticstomorrow.com/story/2022/08/the-developing-role-of-drones-in-manufacturing/19274/

[8] "3D printing drones work like bees to build and repair structures while flying," September 2022, https://penntoday.upenn.edu/news/Penn-Design-Engineering-3D-printing-build-repair-while-flying

[9] "The benefits of drones in manufacturing," June 2022, https://www.themanufacturer.com/articles/the-benefits-of-drones-in-manufacturing/

[10] "Top challenges in drone and UAV manufacturing," May 2024, https://sendcutsend.com/blog/drone-manufacturing-challenges/?srsltid=AfmBOoovaM18IudMnRF7OpigJ4hn3wMS0Ts1NIlRFZhsVss9WK3bGKRM

[11] C. Singh and R. R. Gatti (eds.), *Drone Applications for Industry 5.0*. IGI Global, 2024.

[12] N. K. Singh, P. Muthukrishnan, and S. Sanpini, *Industrial System Engineering for Drones: A Guide with Best Practices for Designing*. Apress, 2019.

CHAPTER 7

DRONES IN CONSTRUCTION

"The whole difference between construction and creation is exactly this: that a thing constructed can only be loved after it is constructed; but a thing created is loved before it exists."

– Charles Dickens

7.1 INTRODUCTION

Drone photos, videos, and imagery are used to scope out projects, track building progress, and provide real-time updates. With their real-time data recording and unique aerial advantage, drones can improve efficiency, cut costs, and streamline workflow. Drones are equipped with multiple features to help capture key data on building sites, including cameras, GPS units, thermal sensors, and infrared sensors.

Drones have quickly become an effective tool for businesses of all sizes. Although many industries are regular users of drone technology, the construction industry has both the largest market share and the fastest-growing market share for drone usage over the past few years. These flying vehicles typically have cameras in them, which means that they provide a birds-eye view of a construction site. With sensors and cameras, drones can also track construction equipment and vehicles on a site.

Drones offer immense possibilities and applications in the world of construction, offering benefits ranging from on-site safety to remote monitoring. They have been making a positive impact on the construction world since becoming a part of the industry years ago. Their current capabilities allow them to cut costs, time, risk, and labor, while improving workflow, accuracy, communication, and efficiency. Whether drones are being used to conduct land surveys or to keep track of equipment, drones are proving invaluable to construction [1].

This chapter explores the applications of drones in the construction industry. It begins with describing what a drone is. It presents some applications of drones in construction. It highlights some benefits and challenges of drones in construction. The last section concludes with comments.

7.2 WHAT IS A DRONE?

The FAA defines drones, also known as unmanned aerial vehicles (UAVs), as any aircraft system without a flight crew onboard. Drones include flying, floating, and other devices, including unmanned aerial vehicles (UAVs), that can fly independently along set routes using an onboard computer or follow commands transmitted remotely by a pilot on the ground. A typical drone is shown in Figure 7.1 [2]. A drone is usually controlled remotely by a human pilot on the ground, as typically shown in Figure 7.2 [2]. Drones can range in size from large military drones to smaller drones. Drones, previously used for military purposes, have started to be used for civilian purposes since the 2000s. Since then, drones have continued to be used in intelligence, aerial surveillance, search and rescue, reconnaissance, and offensive missions as part of the military Internet of things (IoT). Today, drones are used for different purposes such as aerial photography, surveillance, agriculture, entertainment, healthcare, transportation, law enforcement, etc.

Figure 7.1 A typical drone [2].

Figure 7.2 A drone is usually controlled by operators on the ground [2].

Drones work much like other modes of air transportation, such as helicopters and airplanes. When the engine is turned on, it starts up, and the propellers rotate to enable flight. The motors spin the propellers and the propellers push against the air molecules downward, which pulls the drone upwards. Once the drone is flying, it is able to move forward, back, left, and right by spinning each of the propellers at a different speed. Then, the pilot uses the remote control to direct its flight from the ground [3].

Drone laws exist to ensure a high level of safety in the skies, especially near sensitive areas like airports. They also aim to address privacy concerns that arise when camera drones fly in residential areas. These include the requirement to keep your drone within sight at all times when airborne. In the United States, drones weighing less than 250g are exempt from registration with civil aviation authorities. If your drone exceeds 250g in weight, you will also require a Flyer ID, which requires passing a test [4]. It is necessary to register as an operator, be trained as a pilot, and have civil liability insurance, in addition to complying with various flight regulations, and those of the places where their use is permitted.

Most drones have a limited payload, usually under 11 pounds. Drones are classified according to their size. Here are the different drone types:

- Nano Drone: 80-100 mm
- Micro Drone: 100-150 mm
- Small Drone: 150-250 mm

- Medium Drone: 250-400 mm
- Large Drone: 400+ mm

One of the emerging trends in drone use for factories is the utilization of LiDAR technology. LiDAR stands for Light Detection and Ranging. This technology provides accurate depth information essential for understanding the three-dimensional structure of the environment. LiDAR sensors emit laser beams to measure distances to objects, creating high-resolution 3D maps of the surrounding terrain and objects. The ability to capture detailed data through LiDAR technology has opened up opportunities for better predictive maintenance, reduction in inspection times, and overall cost savings [5].

Figure 7.3 A drone in the construction environment [6].

7.3 APPLICATIONS

A drone in the manufacturing environment is shown in Figure 7.3 [6]. While there are many types of drones, commercial drones are the standard commonly used in construction. Drones can be used for a variety of different things throughout the construction industry. They provide construction teams with an overhead view of jobsites, materials, machinery, and people Common applications of drones in construction include the following [7]:

- *Mapping and Land Surveys:* All drone types offer high-quality cameras capable of achieving survey-grade results, with high accuracy. Consulting topographic maps is essential when planning complex, large-scale construction projects. Although topographic maps are helpful, they are expensive and time-consuming to produce. Drones generate topographic map (that shows changes in elevation) with the help of LiDAR lasers. Due to their ability to map vast quantities of land, drones can exponentially cut down on time spent visualizing a site's topography. This allows the construction crew to pinpoint challenges during pre-construction and spot mistakes, saving time and money. A typical land survey is shown in Figure 7.4 [8].

Figure 7.4 A typical land survey [8].

- *Equipment Tracking:* Losing track of where equipment is located at each job site is a common problem. Equipment security is a very important component of managing a job. A manager or drone operator can conduct a flyover to quickly see if a piece of equipment is in a secure enough location and ensure that the equipment is where it should be.
- *Worker Safety:* Worker safety is a top priority of most construction companies. Workers often have to navigate around hazardous conditions when taking manual measurements. Drones can replace workers in these situations and mitigate the risk construction workers face on the field. Construction managers can also use drone video cameras to monitor the job site for safety concerns and ensure that no equipment are loose or unstable. If there is any suspicious activity, drones can send an alert to management to help prevent vandalism or equipment theft.
- *Road Construction:* Building a road is an uphill task which involves a lot of complex engineering and precision work. Every stage of the construction requires careful planning and attention to detail. Drones are making the process a lot easier by providing geographic information system (GIS) data on the road that civil engineers and construction crews can rely on to make more strategic decisions. They also help direct the pouring of asphalt and concrete and provide

detailed visual insight throughout the project. They are a powerful tool for assessing the soil before road construction begins. Figure 7.5 shows the use of a drone in road construction [9].

Figure 7.5 Using a drone in road construction [9].

- *Roof Inspections:* This benefits immensely from using drones. Construction drone photography can eliminate the need for manual roof inspections, which can involve roofs that are wet, slippery, or in dangerously unstable conditions. Drones can get to construction spots, like parts of rooftops that otherwise would be hard and potentially risky to reach. Drones in construction safety can make a big difference in protecting workers.

7.4 BENEFITS

Drones in the construction industry have become commonplace, and the benefits of drones in construction are many. All the drone applications help you avoid making costly mistakes. The tiniest miscalculation spread across miles of road can turn into a big problem. The amount of time and money that drones in construction save is largely substantial. Implementing drones in a construction workflow can completely change the companies' operations throughout the entire bidding, planning, and building process. Through the surveying process, drones can save time and manpower while also mitigating human error. Drones can help with construction projects by improving data collection, enhancing safety, and increasing efficiency. Other benefits of drones in construction include the following [10]:

- *Speed:* One of the main benefits of using drones in construction is their speed. Drones can collect data faster because they are not slowed down by on-the-ground hazards. In the end, road construction surveys that used to take several weeks can now be done in a matter of days with drones.
- *Security:* With large construction sites, the issues of theft, trespassing, and vandalism make monitoring construction sites difficult to manage. To combat this, construction companies use drones to surveillance the area. Along with monitoring construction project progress, drones

can do around-the-clock, real-time job site monitoring. They can protect the site from intruders like unaccompanied visitors or even thieves or vandals.
- *Collaboration:* Drones have allowed construction companies to improve communication and collaboration across their workforce. They can improve internal collaboration for teams by sending information to connected software during flyovers. Design teams, engineers, construction managers, workers, and owners can access the data simultaneously, follow the project, and catch any mistakes that may have occurred.
- *Improved Decision Making:* Drone data helps improve on-the-ground decision-making. A drone can also give decision-makers regular updates and keep everyone informed about a construction's progress. With these insights, those in command of a site can catch pitfalls and mistakes before they balloon into a larger problem.
- *Improved Worker Safety:* Drones increase worker and site safety. Since drones capture data from the air, you do not need to impose unnecessary risk on workers when inspecting the road in or near moving traffic. Overall, drones lower worker and site safety issues to a much more manageable level. This leads to fewer worker accidents, less liability, and lower insurance premiums.
- *Reduced Labor:* Adapting drones into the construction industry lessens the amount of labor and time needed to create accurate surveys. Drones also lessen the risk of human errors and capture crucial data in a fraction of the time as traditional methods. With the help of drones, the project manager can monitor overall productivity, enabling him to keep workers productive.
- *Cost Saving:* For any construction company, there are three main aspects that matter – time, quality and cost. Drones are more affordable than mobilizing a technical inspection team. A construction company's investment in construction drone services is a practical and future-forward decision. The return on investment (ROI) speaks for itself through many business benefits such as increased profit margins, more effective customer acquisition, and staying on schedule.
- *Efficiency:* Anyone who has ever been through the process of manually surveying a construction site can tell you that it is not an easy process. It takes time, a workforce with a very particular skill set, money, and a willingness to redo the survey if anything goes wrong. With drone surveying, even beginner surveyors are capable of surveying 120 acres per hour for a 60x improvement in surveying efficiency.
- *Saving Time*: Drones can help avoid material shortages and overages by calculating stockpile volumes and tracking changes to earthwork. They can also monitor sites for changes that might indicate potential hazards, such as landslips.
- *Fast Data Collection:* Drones are a faster and safer alternative to traditional land-based surveying methods in construction projects. They can complete survey work in 60% to 70% less time and eliminate risks to personnel in rugged terrain.

7.5 CHALLENGES

Any technology tool is only as good as your ability to use it, and drones are no exception. Construction sites do not exist in vacuums; they operate on deadlines and are subject to weather and are themselves rugged environments. Other challenges of drones in construction include the following [11]:

- *Privacy:* Drones can capture images or videos of people or private property, raising privacy concerns.
- *Training and Certification:* Drones may be small, but they are still a type of aircraft. They require training and certification to operate. A nation's aviation authority regulates all unmanned aerial vehicles (or UAVs). The location of the construction worksite can determine how, when, and if you fly, depending on the airspace overhead.
- *Regulations:* Flying drones is highly regulated. If you are close to an airport or other restricted airspace, you may find that you cannot legally fly a drone. Most countries have different sets of regulations for drones. US commercial drone regulations are managed by the FAA.
- *Injury:* When drones veer off course, they can hit people and their sharp blades can cause injury and death. Construction projects can be dangerous. Construction worker fatalities in 2019 represented over 20% of all on-the-job deaths, making construction one of America's most dangerous industries. A significant portion of construction project costs are injury related. This leads to increased insurance prices and reduced profit margins. Injury may cost lost project time, litigation, medical bills, and compensation. A deadly accident can end a construction company all together.
- *Hacking:* Like most connected devices, drones are vulnerable to hacking. Hackers can quickly attack the central control systems of some commercial drones from a mile away to access private information, destroy or damage files, and leak data to unauthorized third parties.
- *Weather:* Drones are more sensitive to weather conditions than conventional aircraft, and can be blown off course or damaged by wind, rain, or hail. They can be unsafe to fly in bad weather, such as strong winds, rain, or extreme temperatures. This can delay projects and affect the accuracy of data collection.
- *Battery Life*: Most commercial drones have a flight time of 20–30 minutes per charge, which may not be enough for long-duration flights or monitoring large areas.
- *Safety:* Drones can pose a risk of collision with workers or other construction equipment. They can also cause distraction, which can increase the likelihood of falls.
- *Liability:* Construction companies can be held liable for accidents or property damage caused by drones. To mitigate this risk, companies should get insurance that specifically covers drone operations.
- *Lack of Resources*: Some construction companies may not have the expertise or resources to use drones effectively.

7.6 CONCLUSION

The use of drones in construction and construction-related industries has now become commonplace, with construction being the fastest-growing commercial industry adopting drones. Drones have become the go-to tool for construction firms to track, map, survey, inspect, and manage worksites more efficiently and safely. They are increasingly becoming pivotal in the construction industry, bringing unparalleled value in terms of efficiency, safety, and accuracy. For drone pilots looking to launch a new career in commercial drone flight, the construction industry is a field worth consideration.

The impact of drones is so great that companies will fall behind industry standards without them. While the impact of drones in the construction industry is evolving, they are already revolutionizing the sector and changing the way construction projects are handled. As technology advances, drones will be able to take on more construction tasks than ever before. Drones will continue to offer industry leading solutions that deliver the highest quality of speed, accuracy, and cost effectiveness. It is safe to say that the commercial drone is here to stay. More information about drones in construction can be found in the books in [12-13 14 15 16].

REFERENCES

[1] M. N. O. Sadiku, M. Oteniya, J. O. Sadiku, and S. Abunene, "Drones in construction," *International Journal of Scientific and Research Publication*, vol. 14, no. 9, September 2024, pp. 285-294.

[2] "Drone in construction & infrastructure," https://www.jouav.com/industry/drone-in-construction

[3] "How drones work and how to fly them," May 2024, https://dronelaunchacademy.com/resources/how-do-drones-work/

[4] "What are the main applications of drones?" June 2024, https://www.jouav.com/blog/applications-of-drones.html

[5] "Drones in manufacturing: A game-changer for industry," https://viper-drones.com/industries/infrastructure-drone-use/manufacturing/#:~:text=The%20integration%20of%20drones%20into,on%20manufacturing%20is%20no%20exception.

[6] J. Moraglia, "Benefits of using a drone for construction," January 2024, https://thedronelifenj.com/benefits-of-using-a-drone-for-construction/

[7] L. Stannard, "6 Ways drones in construction are changing the industry," February 2022, https://www.bigrentz.com/blog/drones-construction?srsltid=AfmBOooXvjQzNrIfYOlUib380fStZaXLZCBA8mN8C18PzaYyy79ysTHP

[8] "Six ways drones in construction add value," https://wingtra.com/drone-mapping-applications/drones-in-construction-and-infrastructure/

[9] "Drones in construction – Why they are beneficial and how to use them," https://www.propelleraero.com/blog/drones-in-construction-why-they-are-beneficial-and-how-to-use-them/

[10] "Benefits of drones in construction and infrastructure projects," https://www.jouav.com/industry/drone-in-construction

[11] "4 Disadvantages of drones in construction," https://timelapselab.it/en/construction-site-monitoring-blog/4-disadvantages-of-drones-in-construction.html

[12] D. Tal and J. Altschuld, *Drone Technology in Architecture, Engineering and Construction: A Strategic Guide to Unmanned Aerial Vehicle Operation and Implementation.* Wiley, 2021.

[13] F. D'Alessandro, *Drones in Construction Sites.* CreateSpace Independent Publishing Platform, 2015.

[14] L. B. P. Mamani, *Making Money with Drones, Drones in the Construction Industry.* Independently Published, 2021. ·

[15] J. Powell, *Drone Services in Construction.* Office of Industries, US International Trade Commission, 2021.

[16] B. Hoffstadt, *Success with Drones in Civil Engineering: An Accelerated Guide to Safe, Legal, and Profitable Operations.* Independently Published, 2018.

CHAPTER 8

DRONES IN OIL & GAS INDUSTRY

"The good Lord didn't see fit to put oil and gas only where there are democratic regimes friendly to the United States."

- Dick Cheney

8.1 INTRODUCTION

Drones have proven themselves a unique and powerful asset through all their work. As a result, all industries are using them today to get various tasks done. Industries like e-commerce, defense, entertainment, manufacturing, telecommunications, etc. have been using drones since their early development. The application of drone technology has transformed many industries, including the oil and gas industry. Drones have become a crucial part of the oil and gas industry, where they are being deployed to perform a wide variety of tasks. Drones in oil and gas have helped reduce inspection time, cut costs, decrease downtime, and identify problems early on. When it comes to oil and gas industry, drones are a super significant investment. They have made it possible to automate inspections, ease management, and save lives [1]

The oil and gas industry, known for its vast operations and extensive infrastructure, is undergoing a technological renaissance with the integration of drone inspections. An example of such extensive infrastructure is shown in Figure 8.1 [2]. The oil and gas sector has long been at the forefront of adopting cutting-edge technologies to enhance safety, efficiency, and cost-effectiveness. Drones are transforming the oil and gas industry, offering a range of applications that enhance operational efficiency, safety, and environmental compliance. By leveraging the capabilities of drone technology, oil and gas companies can streamline operations, reduce costs, improve safety outcomes, and make informed decisions based on real-time data.

Figure 8.1 An extensive infrastructure for oil and gas plant [2].

Drones are transforming oil and gas exploration and providing numerous benefits for the companies that use them. They enable companies to optimize their operations, improve safety, and minimize environmental risks, making them indispensable in the oil and gas industry. They offer a cost-effective solution for routine inspections and data collection, reducing the need for extensive manpower and helicopter-based surveillance. One driving factor behind the remarkable oil and gas drones market growth is the cost-effectiveness and efficiency they bring to the industry. Consequently, cost-conscious companies increasingly turn to drones to optimize their operations, making this a pivotal driver of market expansion [3].

This chapter examines the applications of drones in the oil and gas industry. It begins with describing what a drone is. It covers oil and gas drones. It presents some applications of drones in the oil and gas industry. It highlights some benefits and challenges of drones in the oil and gas industry. The last section concludes with comments.

8.2 WHAT IS A DRONE?

The FAA defines drones, also known as unmanned aerial vehicles (UAVs), as any aircraft system without a flight crew onboard. Drones include flying, floating, and other devices, including unmanned aerial vehicles (UAVs), that can fly independently along set routes using an onboard computer or follow commands transmitted remotely by a pilot on the ground. A typical drone is shown in Figure 8.2 [4].

A drone is usually controlled remotely by a human pilot on the ground, as typically shown in Figure 8.3 [5]. Drones can range in size from large military drones to smaller drones. Drones, previously used for military purposes, have started to be used for civilian purposes since the 2000s. Since then, drones have continued to be used in intelligence, aerial surveillance, search and rescue, reconnaissance, and offensive missions as part of the military Internet of things (IoT). Today, drones are used for different purposes such as aerial photography, surveillance, agriculture, entertainment, healthcare, transportation, law enforcement, etc.

Figure 8.2 A typical drone [4].

Figure 8.3 A drone is usually controlled by operators on the ground [5].

Commercial drones have come a long way in the last decade. Drones work much like other modes of air transportation, such as helicopters and airplanes. When the engine is turned on, it starts up, and the propellers rotate to enable flight. The motors spin the propellers and the propellers push against the air molecules downward, which pulls the drone upwards. Once the drone is flying, it is able to move forward, back, left, and right by spinning each of the propellers at a different speed. Then, the pilot uses the remote control to direct its flight from the ground [6].

Drone laws exist to ensure a high level of safety in the skies, especially near sensitive areas like airports. They also aim to address privacy concerns that arise when camera drones fly in residential areas. These include the requirement to keep your drone within sight at all times when airborne. In the United States, drones weighing less than 250g are exempt from registration with civil aviation authorities. If your drone exceeds 250g in weight, you will also require a Flyer ID, which requires passing a test [7]. It is necessary to register as an operator, be trained as a pilot, and have civil liability insurance, in addition to complying with various flight regulations, and those of the places where their use is permitted.

Most drones have a limited payload, usually under 11 pounds. Drones are classified according to their size. Here are the different drone types:

- Nano Drone: 80-100 mm
- Micro Drone: 100-150 mm
- Small Drone: 150-250 mm
- Medium Drone: 250-400 mm
- Large Drone: 400+ mm

One of the emerging trends in drone use for factories is the utilization of LiDAR technology. LiDAR stands for Light Detection and Ranging. This technology provides accurate depth information essential for understanding the three-dimensional structure of the environment. LiDAR sensors emit laser beams to measure distances to objects, creating high-resolution 3D maps of the surrounding terrain and objects. The ability to capture detailed data through LiDAR technology has opened up opportunities for better predictive maintenance, reduction in inspection times, and overall cost savings [8].

8.3 OIL&GAS DRONES

The oil and gas industry is one of the biggest industries in the world. It is one of the most capital-intensive while being one of the largest contributors to the global economy. Oil and gas firms are regarded as one of critical industries. The oil and gas industry has traditionally been a conservative sector, hesitant to embrace new technologies. However, the advent of drones has provided the industry with new opportunities to optimize operations, reduce costs, and improve safety. In the last decade, oil and gas companies have harnessed the mobility and perspective of drones to innovate workflows and save money.

North America, particularly the US, leads the world in drone adoption, while European markets follow closely. In 2013, BP became the first oil and gas company to receive a license to operate drones. Several oil and gas companies such as ExxonMobil, Shell, and Chevron followed suit. The oil and gas drones market size is experiencing robust growth due to several key factors. There is a growing emphasis on cost reduction and operational efficiency in the energy sector.

There are drones specifically designed for the oil and gas industry. Oil and gas drones encompass a range of specialized UAV technologies tailored for the oil and gas industry. These drones are equipped with high-resolution cameras, sensors, and data analysis software. The drones provide crucial data for inspections, maintenance, and exploration. The drones can be used to monitor pipelines, check for leaks, and conduct surveys of oil spill zones. They can also fly over rough terrain, making them ideal for inspections of pipelines and other infrastructure placed in unmanageable terrain. The drone technology allows for increased efficiency, improved safety, access to remote areas, cost-effectiveness, enhanced data collection, and improved response to accidents [9]. A growing variety of sensors allow oil and gas drones to perform more tasks and make these tasks more sophisticated. Today, the oil and gas industry cannot do without drones. A typical oil and gas drone is displayed in Figure 8.4 [5].

Figure 8.4 A typical oil and gas drone [5].

8.4 APPLICATIONS

Drone deployment has significantly transformed the oil and gas industry, especially for operations in difficult-to-reach areas such as offshore rigs and pipelines. Drones have been applied in diverse capacities ranging from drone inspection, asset inspection, leak detection, pipeline inspection, flare stack inspection, and emergency response, etc. As more and more commercial drone manufacturers work closely with oil and gas companies around the world, more customized applications are made available. Advancements in drone technology are also increasing its applications. Common applications of drones in oil and gas sector include the following:

- *Inspection:* Oil & gas inspection applications are at the forefront of drone adoption in the industry. Drones can collect better data than humans can because they can get much closer to infrastructure. The drones are deployed for critical tasks such as pipeline monitoring, rig inspections, and facility surveillance, enhancing safety, and operational efficiency. While environmental impact assessment and other applications are significant, it is the pressing need for infrastructure inspection and maintenance that drives the dominance of the oil and gas inspection segment. Drones can run inspections without needing to shut down oil operations. They help you save on several maintenance costs, which in turn frees up funds to conduct more frequent drone inspections. An example of drone for oil and gas inspection is shown in Figure 8.5 [10].

Figure 8.5 A drone for oil and gas inspection [10].

- *Flare Stack Inspection*: Flare stacks, used to dispose of waste gases safely, require regular inspections to ensure their integrity and functionality. Traditionally, manual inspections of flare stacks involve significant downtime and safety risks for workers. Drones can capture detailed images, assess heat signatures, and detect potential issues, enabling proactive maintenance and minimizing downtime. Figure 8.6 shows flare stacks in operation [11].

Figure 8.6 Flare stacks in operation [11].

- *Leak Detection:* Before drones, oil and gas companies tried to detect leaks by mounting fixed detectors at high-risk spots in facilities and along pipelines or by having inspectors occasionally check areas with portable detectors. However, these traditional leak detection methods can be costly and inefficient. Drones equipped with gas detectors can quickly identify the location of leaks and help determine the extent of the leak. Drone-based leak detection is faster and more comprehensive. Drones can easily detect spills and corrosion in flare stacks that are high off the ground. Drones equipped with sensors and artificial intelligence can detect gas leaks, pipeline corrosion, and crude oil spills, enabling emergency response teams to take action quickly.
- *Environmental Monitoring:* Drones can now be equipped with multispectral and hyperspectral cameras and are used to monitor the environmental impact of oil and gas operations. They can assess soil health, vegetation health, and water quality in and around drilling sites. This way, companies can adhere to environmental regulations, ensure responsible resource extraction, and minimize ecological disruption.
- *Security:* Drones add an extra layer to field security by patrolling expansive oil fields, pipelines, and remote facilities. Drones can patrol remote facilities, pipelines, and oil fields to detect unauthorized activities, trespassing, and security breaches. They provide real-time visual feeds to security personnel, enabling rapid responses and enhancing overall site security. They can be really helpful in identifying and mitigating safety risks. They can be a great tool for managing risks by reducing human exposure to hazardous environments, drones enhance worker safety, and support adherence to strict safety protocols.
- *Data Collection:* Drones can provide high-quality data in a fraction of the time and at a lower risk to humans. Drones can be equipped with advanced sensors to gather massive amounts of data during flyovers. This data is then processed and analyzed using specialized software. The

result guides reservoir management decisions, optimize drilling techniques, and improve overall production efficiency. The incorporation of AI and data analytics will make data collection and analysis quicker, more accurate, and more comprehensive.
- *Monitoring:* The need for autonomous monitoring and inspection in the oil and gas industry is clear. Oil and gas companies use drones to remotely check and observe equipment, infrastructure components, trucks, tankers, and other company assets. Drones help monitor and manage long pipelines spread across different geographies, identifying leakages or anomalies quickly. The drones are able to deliver 360-degree views of subjects for the monitoring of field operations, keep an eye on the development of new facilities, and detect encroachment on pipelines, railways, and other valuable company property. Drone-powered inspections and safety monitoring improve efficiency without halting operations or compromising personnel safety.

8.5 BENEFITS

The use of drones in oil and gas industry has several benefits that make exploration safer, more efficient, and cost-effective. Drones can save millions of dollars in asset maintenance costs. They also allow you to cut down on labor costs. Their combination of safety, efficiency, precision, and cost-effectiveness makes them an indispensable tool in the modern oil and gas inspection toolkit. Other benefits of drones in oil and gas industry include the following [2,12]:

- *Safer & Faster Inspections:* Traditional oil inspections are dangerous to say the least. But drones are making inspections much safer. These traditional inspection methods can be costly and inefficient. Drones inspect better because of their advanced cameras. They can fly to great heights and through toxic chemicals with ease, so there is no need to put personnel in harm's way. While traditional oil and gas inspections can take several days to complete, drones fly directly to a target at command, perform comprehensive checks, and eliminate the need to shut down operations. Nearly every stage of petroleum production benefits from close-eyed inspections by drones.
- *Increased Efficiency:* Drones can collect data at a much faster pace than traditional methods. They can perform routine inspections faster and provide better visibility of assets. This enables oil and gas companies to survey large areas quickly and allows them to detect faults and geographical features at a much faster pace. All of this can be done in a short amount of time and at a reduced cost because drones do not need as much manpower as traditional methods require. Drones have the capacity to reduce operational expenses significantly by eliminating the need for extensive human labor and expensive helicopter-based surveillance.
- *Improved Safety:* Safety is a significant concern in all aspects of the oil and gas sector. In fact, a significant benefit of drone technology is safety. Traditional methods of oil and gas exploration often involve on-site staff who are at risk of physical harm and exposure to hazardous substances. By using drones to conduct surveys and inspections, oil and gas companies can avoid putting their staff in danger. Drones can eliminate traditional methods of inspecting areas like essential production components in oil refineries, chimneys, smokestacks, storage tanks, jetties, and other potentially hazardous environments.

- *Access to Remote Areas:* Some oil and gas exploration takes place in remote and inaccessible areas. By using drones, oil and gas companies can survey these areas without putting their staff in danger.
- *Cost-Effectiveness:* Drone technology is more cost-effective than traditional methods of oil and gas exploration. By using drones, oil and gas companies can reduce their expenses while still collecting valuable data that can inform their operations. The cost savings extend to safety enhancements, as drones can access hazardous or remote locations without endangering human lives. Drones can be used to inspect thousands of miles of pipelines at a fraction of the cost and at a more precise level.
- *Enhanced Data Collection:* Drone technology is transforming data collection and analysis in the oil and gas industry, allowing greater accuracy and speed. Drones are equipped with sensors that can measure heat, pressure, and humidity levels, which provides real-time data about natural gas and oil reserves.
- *Improved Response to Accidents*: In the event of an accident in the oil and gas industry, time is of the essence. Drones can help oil and gas companies respond more quickly to accidents and provide more accurate information about the extent of the damage.
- *Environmental Regulation:* Stringent safety and environmental regulations are pushing oil and gas companies to adopt advanced technologies for monitoring and maintaining their infrastructure. Stricter environmental regulations and heightened concerns over the ecological impact of the industry have pushed energy companies to adopt more responsible practices.
- *Industry Giants:* The oil and gas drones market is significantly influenced by key industry giants that play a pivotal role in driving market dynamics and shaping consumer preferences. Their strong global presence and brand recognition have contributed to increased consumer trust and loyalty, driving product adoption. These industry giants continually invest in research and development, introducing innovative designs, materials, and smart features.
- *Emergency Response:* One of the key challenges in emergency situations is getting accurate information to the decision-makers in real time. It is important to have a strategy in place when disasters happen. Disasters may be oil spill, natural, mechanical blowout, etc. Drones can help you respond quickly and efficiently. Unlike manned aircraft that need a pilot and lead time to be ready for takeoff, drones can launch immediately. Their speedy deployment can save precious time when disaster hits. Drones can also ensure emergency response equipment is set up in the right spots.
- *Deliveries:* Drones can carry out deliveries of material, replacement parts, and supplies to offshore rigs and other remote operations, instead of sending out a boat or helicopter to reach the rig.
- *Real-Time Communication:* Communication is crucial during exploration, inspection, production, and other processes, as there should be clear and prompt data transmission between professionals and departments involved. Proceeding with oil and gas drone inspection also secures real-time communication. The drone pilot can capture and record the situation and directly send it to authorized offices for viewing and analysis.
- *Accurate Data Gathering:* Manual oil and gas topside corrosion diagnoses and inspections are manual activities prone to many bottlenecks, and prone to inaccuracy. What makes drone technology an invention worth the investment is that it secures accurate data that can help in the success of the work operation. Drone manufacturers are continuously improving their products by meeting the needs of different industries. The best thing to expect with drones

is the ability to record and present precise and detailed imaging data that can help track and analyze the work process.
- *Environmental Impact:* The oil and gas sector is under increasing pressure to minimize environmental impact. Drones offer a more eco-friendly alternative to conventional methods. They operate silently, minimize ground disturbances, and leave a reduced carbon footprint. Their ability to operate without disturbing local ecosystems makes them an environmentally responsible choice.
- *Accessibility:* There are areas within the vast landscapes of the oil and gas sector that are incredibly hard to reach, or even inaccessible through traditional means. Drones effortlessly navigate these areas, ensuring that no asset remains unchecked.
- *Continuous Monitoring:* Drones enable 24/7 monitoring of assets and infrastructure. Drones can be programmed to conduct scheduled or on-demand inspections, capturing high-resolution images and videos for analysis. Drones help monitor and manage long pipelines spread across different geographies, as typically shown in Figure 7 [5].

Figure 7 Drones help monitor and manage long pipelines [5].

8.6 CHALLENGES

The oil and gas industry faces numerous challenges, such as remote and harsh operating environments, high safety risks, the need for cost-effective solutions, and the inherent risks associated with operations. Other challenges of drones in oil and gas industry include the following [10]:

- *Complex Regulation:* One significant challenge facing oil and gas drone is the complex regulatory environment. Drones operate in an airspace that is governed by a myriad of regulations. In many regions, there are restrictions on flight altitudes, no-fly zones, and specific permissions required for commercial use. Companies must adhere to various national and international regulations concerning drone usage, such as registration, flight restrictions, and operator certifications. The use of drones in the energy sector is subject to strict aviation regulations, which can vary by region and country. Ensuring drones adhere to both aviation and industry-specific guidelines adds complexity to their deployment. Overcoming these regulatory and compliance hurdles is a challenge for market players, which can impede the seamless integration of drones into oil and gas operations. There is good news for companies that build drone applications for oil and gas: the regulatory climate seems to be finally starting to relax.
- *Safety:* One of the paramount concerns in the oil and gas sector is ensuring the safety of its personnel. Employee safety is of paramount importance in refineries, which are inherently hazardous environments. There are safety concerns such as drones entering restricted airspace, collisions with other aircraft, and human error that can impact operations. With emissions regulation tightening and public awareness of environmental issues growing, the oil and gas industry is under increasing pressure to take safety seriously. Drones eliminate direct human exposures, allowing for inspections in even the most challenging and hazardous environments without jeopardizing human safety.
- *Privacy:* Concerns related to privacy and data security further complicate their widespread adoption.
- *Cost:* Purchasing drones can be expensive, considering that they will also require a considerable investment in terms of drone pilots, training, and maintenance.
- *Expertise Required:* Drone operation and analysis require specific skills and expertise that can be a challenge for the oil and gas industry. Operating a drone, especially in complex environments like oil and gas facilities, requires specialized skills. Companies need to train drone pilots who can operate the vehicles safely, follow flight plans, and analyze the data collected. Additionally, they need data scientists who can analyze the information and provide insights that are actionable.
- *Technical Limitations:* In spite of their advanced capabilities, drones do have limitations. The range of drones can be limited, implying that they can only operate in a specific area, limiting the extent to which they can be used to replace manual inspection and monitoring activities. Battery life can restrict the duration of inspection flights, which is reduced by cold temperatures. Drones also face challenges operating in extremely windy and rainy weather conditions, limiting their operational days.
- *Data Management:* In the vast world of oil and gas inspections, drones have emerged as technological marvels, capturing massive amounts of detailed data. A fundamental challenge with drone inspections is the sheer volume of data collected. Drones can capture massive amounts of data in a single flight. While this is an advantage, it can also lead to information overload that needs effective storage and analysis. Without an efficient system or platform to manage this data, crucial insights might be overlooked, or the data may simply be inaccessible to those who need it.
- *Integration with Existing Systems:* Companies in the oil and gas sector often have legacy systems in place. Integrating drone-captured data with these existing systems can be challenging and might require additional software or platform solutions.

- *Environmental Concerns:* While drones are relatively less invasive, they can sometimes raise concerns related to noise, privacy, and potential disturbances to local wildlife. Ensuring that drone operations are environmentally sensitive and respectful of local communities is essential.
- *Climate:* Harsh weather conditions such as low temperatures, wind, precipitation, fog, and cloud cover can make it difficult for drones to operate. Electronics may not be designed for very low temperatures, which can cause drones to lose airborne control.

8.7 CONCLUSION

Specialists in many industries are exploiting the unique flexibility and observational capabilities of drones to improve industrial processes and operational efficiency. Drones have come into use in the oil and gas industry as a precise, highly maneuverable, and cost-efficient means of carrying out transportation network and asset inspections. The use of drones has significantly reduced the risk to personnel and improved the efficiency of inspections.

The adoption of drone technology will become increasingly common throughout the oil and gas industry as companies discover the advantages of using drones. As the industry moves forward, it is evident that the synergy between drone technology and advanced data management platforms will define the future of oil and gas inspections. Oil and gas companies that embrace drone technology will have a significant advantage. One major trend in the oil and gas drones is the integration of artificial intelligence (AI) and machine learning. Drones equipped with advanced AI algorithms can autonomously detect anomalies in infrastructure, predict maintenance needs, and optimize operational processes. The future of drones in the oil and gas industry looks bright, with advancements in technology continually expanding their capabilities.

REFERENCES

[1] G. Goyal, "Why is drone usage growing exponentially in oil and gas industry?" September 2023, https://www.iotforall.com/why-is-drone-usage-growing-exponentially-in-oil-and-gas-industry#:~:text=Drones%20play%20a%20crucial%20role,or%20corrosion%20along%20pipeline%20surfaces.

[2] "5 Major benefits of drones in oil and gas," https://thedronelifenj.com/benefits-of-drones-in-oil-and-gas/

[3] M. N. O. Sadiku, O. D. Olaleye, and J. O. Sadiku, "Drones in oil & gas industry," *International Journal of Trend in Research and Development*, vol. 11, no. 5, September 2024, pp. 96-101.

[4] "How are drones used in the oil and gas industry?" July 2021, https://copas.org/how-are-drones-used-in-the-oil-and-gas-industry/

[5] "Oil & gas," https://enterprise.dji.com/oil-and-gas

[6] "How drones work and how to fly them," May 2024, https://dronelaunchacademy.com/resources/how-do-drones-work/

[7] "What are the main applications of drones?" June 2024,

      https://www.jouav.com/blog/applications-of-drones.html

[8] "Drones in manufacturing: A game-changer for industry,"
https://viper-drones.com/industries/infrastructure-drone-use/manufacturing/#:~:text=The%20integration%20of%20drones%20into,on%20manufacturing%20is%20no%20exception.

[9] "Oil and gas drones market," August 2024,
https://www.businessresearchinsights.com/market-reports/oil-and-gas-drones-market-109938

[10] "Oil and gas drone inspections: Accelerating asset management," September 2023,
https://optelos.com/oil-and-gas-drone-inspection/

[11] "Revolutionising the oil and gas sector: The rise of drone-in-a-box systems," May 2023,
https://www.linkedin.com/pulse/revolutionising-oil-gas-sector-rise-drone-in-a-box

[12] "6 Unique advantages of drone technology in oil and gas exploration,"
https://www.linkedin.com/pulse/6-unique-advantages-drone-technology-oil-gas-ryan-shore#:~:text=The%20technology%20allows%20for%20increased,and%20improved%20response%20to%20accidents.

CHAPTER 9
DRONES IN POWER SYSTEMS

"People always fear change. People feared electricity when it was invented, didn't they? People feared coal, they feared gas-powered engines... There will always be ignorance, and ignorance leads to fear. But with time, people will come to accept their silicon masters."

– Bill Gates

9.1 INTRODUCTION

The power sector is one of the world's most rigid industries. It is now in the midst of a profound transition. Disrupted not just by economic and environmental forces, the industry is facing rapid technological changes that have forced companies to reevaluate business models to stay profitable. The interest in drone by the power industry is soaring. The reason drones are being adopted so quickly in the power sector is because they make good business sense and for their multitasking capabilities [1]. The task of maintaining and inspecting high voltage transmission and distribution lines can be difficult, dangerous, and expensive. As a result, utilities are increasingly using drones as a safe and effective tool to assist them in their operations. Drones drastically cut the costs of power line inspections for utilities. They also improve safety, increase reliability, and reduce response time across transmission and distribution systems.

Commercial drones have come a long way in the last decade. The use of drones, in the power industry is relatively new, with many companies still unsure about the value the drones could bring to their operations. Drone technology can assist engineers in the process of designing the electrical infrastructure. It can drastically reduce asset inspection time and save labor costs, while providing higher-quality data that enables companies to maximize energy production. Today's energy and utilities companies are using drones to capture data that was previously too dangerous, difficult or expensive to obtain. National grid is also actively deploying drones to help improve power line inspections, especially fault response in rural power grids, help restore power supply, and improve the safety, efficiency, and reliability of power systems.

Modern drone technology has practically transformed the landscape of every industry. While numerous tools and methods already exist for power line inspection and maintenance tasks, the drones are slowly proving that they offer a practical and valuable approach. Drones are increasingly used in the power industry. They make it much easier for us to inspect the overhead wires and the national grid, which transmit electricity from power stations across the nation. They can improve efficiency, speed, safety, and cost from the design, pre-construction, and development stages through to commissioning and ongoing maintenance [2].

This chapter examines the applications of drones in power systems. It begins with describing what a drone is. It presents some applications of drones in power systems. It highlights some benefits and challenges of drones in power systems. The last section concludes with comments.

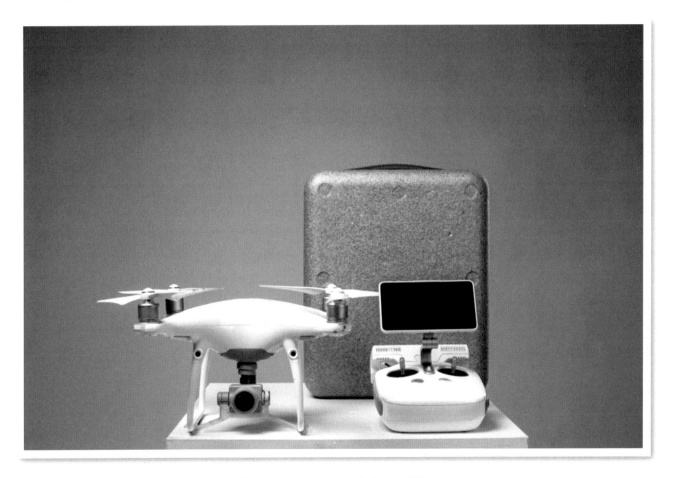

Figure 9.1 A typical drone [3].

9.2 WHAT IS A DRONE?

The FAA defines drones, also known as unmanned aerial vehicles (UAVs), as any aircraft system without a flight crew onboard. Drones include flying, floating, and other devices, including unmanned aerial vehicles (UAVs), that can fly independently along set routes using an onboard computer or follow commands transmitted remotely by a pilot on the ground. A typical drone is shown in Figure 9.1 [3]. A drone is usually controlled remotely by a human pilot on the ground, as typically shown in Figure 9.2 [4]. Drones can range in size from large military drones to smaller drones. Drones, previously used for military purposes, have started to be used for civilian purposes since the 2000s. Since then, drones have continued to be used in intelligence, aerial surveillance, search and rescue, reconnaissance, and offensive missions as part of the military Internet of things (IoT). Today, drones are used for different purposes such as aerial photography, surveillance, agriculture, entertainment, healthcare, transportation, law enforcement, etc.

Figure 9.2 A drone is usually controlled by operators on the ground [4].

Commercial drones have come a long way in the last decade. Drones work much like other modes of air transportation, such as helicopters and airplanes. When the engine is turned on, it starts up, and the propellers rotate to enable flight. The motors spin the propellers and the propellers push against the air molecules downward, which pulls the drone upwards. Once the drone is flying, it is able to move forward, back, left, and right by spinning each of the propellers at a different speed. Then, the pilot uses the remote control to direct its flight from the ground [5].

Drone laws exist to ensure a high level of safety in the skies, especially near sensitive areas like airports. They also aim to address privacy concerns that arise when camera drones fly in residential areas. These include the requirement to keep your drone within sight at all times when airborne. In the United States, drones weighing less than 250g are exempt from registration with civil aviation authorities. If your drone exceeds 250g in weight, you will also require a Flyer ID, which requires passing a test [6]. It is necessary to register as an operator, be trained as a pilot, and have civil liability insurance, in addition to complying with various flight regulations, and those of the places where their use is permitted.

Most drones have a limited payload, usually under 11 pounds. Drones are classified according to their size. Here are the different drone types:

- Nano Drone: 80-100 mm
- Micro Drone: 100-150 mm
- Small Drone: 150-250 mm
- Medium Drone: 250-400 mm
- Large Drone: 400+ mm

One of the emerging trends in drone use for factories is the utilization of LiDAR technology. LiDAR stands for Light Detection and Ranging. This technology provides accurate depth information essential for understanding the three-dimensional structure of the environment. LiDAR sensors emit laser beams to measure distances to objects, creating high-resolution 3D maps of the surrounding terrain and objects. The ability to capture detailed data through LiDAR technology has opened up opportunities for better predictive maintenance, reduction in inspection times, and overall cost savings [6].

Figure 9.3 A drone in the power system environment [8].

9.3 APPLICATIONS

A drone in the power system environment is shown in Figure 9.3 [8]. Drones are used in power systems for a variety of applications. Common applications of drones in power systems include the following [4,8]:

- *Inspecting Power Lines:* Companies in almost every industry use drones for inspections, as it is a cost-conscious and effective way to inspect at heights and inaccessible areas. Drones can inspect parts of a network that are difficult or expensive to reach, such as overhead wires, pylons, transmission towers, and substations. Inspectors usually conduct power line and utility pole inspections on foot, by car, or in a helicopter, with most inspections taking place on the ground. For rural power lines and utility poles, rough terrain makes walking or driving challenging and incredibly slow. Drones offer a cleaner, cheaper option for power line and utility pole inspections. By performing frequent drone inspections, companies save time, keep the environment and employees safe, and ensure that assets function efficiently. The 3-D images obtained give inspectors a highly accurate snapshot of power lines and poles. Compared to traditional manual inspection, the main advantage of drone inspection lies in greater quality and detail in the inspection. Figure 9.4 shows line inspection with drones [8].

Figure 9.4 Line inspection with drones [8].

- *Power Line Mapping*: Drones can map thousands of kilometers of electrical lines to help optimize maintenance and monitoring plans. This can include monitoring vegetation, verifying maintenance work, and checking the general condition of the power lines.
- *Overhead Lines:* Drones are expected to become increasingly autonomous, with enhanced real-time decision-making and problem-solving capabilities. Automatic inspection of power lines has become a big trend. Previously, overhead cable lines were only repaired after problems occurred. Today, with drone supervision, many problems can be eradicated in their initial phase. Drones automatically take off from the substation, carry out autonomous inspections strictly in accordance with the predetermined inspection area, and complete on-site inspection. Drones

can fly close to the power line and send real-time data to the operators. Autonomous inspection with drones has achieved point-to-point and point-to-line inspections, and is also significantly reducing the difficulty of operation for personnel and the risk of damage to drones. A drone for inspecting the national grid is shown in Figure 9.5 [9].

Figure 9.5 A drone for inspecting the national grid [9].

- *Transmission Line Maintenance:* Factors like environmental conditions, vegetation encroachment, and structural deterioration pose risks to the integrity of these lines, leading to power outages and often safety hazards. Proactive, or preventative, maintenance can help to mitigate these incidents. Transmission line safety is critical for maintaining an uninterrupted and reliable power supply. Drones have revolutionized the way maintenance inspections are conducted for transmission lines. Drones with high-definition cameras can fly close to power lines and take high-resolution images, allowing operators to identify any problems on the wires, such as breaks, wear and tear, weather damage, etc. They automate inspections and provide accurate data on all parts of the power grid, enabling timely repairs, removing risk of accidents, and reducing downtime. Drones enhance transmission line safety through proactive maintenance. Figure 9.6 displays a drone used for power line maintenance [10].

Figure 9.6 A drone for power line maintenance [10].

- *Integration with Smart Grids*: Drone inspection will be further synchronized with smart grid operations. Drones will not only collect data, but they will also interact with the network infrastructure in real time, facilitating more proactive maintenance and faster response to network fluctuations or detected problems.
- *Renewable Energy:* Renewable energy production contributed 18.5 percent of the earth's total electricity in 2018. This was mainly generated from the hydroelectric power systems. Although the use of wind energy is not new, the introduction of drones to help in electricity production has given it a new dimension.
- *Corona Discharge*: Corona discharge is a luminous partial discharge from conductors and insulators due to ionization of the air, where the electrical field exceeds a critical value. Corona discharges can generate corrosive materials, like ozone and nitrogen oxides. Utilities are typically made aware of corona by complaints of faulty radio or television signals. Because corona are invisible in daylight with the naked eye, maintenance crews will investigate by aiming devices such a corona camera or radio antenna at suspected areas, and track corona. Drones equipped with corona cameras can detect corona discharges on power lines from a safe distance and generate real-time images of problem areas.

9.4 BENEFITS

Drones drastically cut the costs of power line inspections for utilities. They also improve safety, increase reliability, and reduce response time across transmission and distribution systems. Drones in the power line industry have been thriving due to the benefits it brings to the table. With these benefits, it is easy to understand the growing attraction to drones in power systems. Other benefits of drones in power systems include the following [9,11]:

- *Efficiency:* One of the primary benefits of utilizing drones for power line inspections is the significant cost-efficiency versus traditional inspection methods. Drone inspection not only allows inspection work to be completed quickly and accurately in a short period of time, but also prevents the risk of injury from falls to personnel, greatly increasing work efficiency and safety coefficient. Drones significantly enhance maintenance efficiency by streamlining the inspection process.
- *Automation:* By establishing three-dimensional models of power lines and improving databases, drones expected to be able to perform automatic energy inspections.
- *Decision-Making:* Drones equipped with advanced sensors and AI algorithms provide utility companies with accurate and real-time data. With this data maintenance schedules can be optimized based on the condition of transmission lines, prioritizing areas that require immediate attention and efficiently allocating resources.
- *Cost Savings:* Manual inspections involve significant costs associated with labor, equipment, and logistics. Drones equipped with high-resolution cameras and advanced imaging technologies efficiently capture detailed visual data of the power lines at a fraction of the cost. These cost savings can be significant in the long run.
- *Time Saving:* Another significant benefit drones have is the substantial time savings they offer through efficient data collection and faster overall inspections. Drones provide a safe way to get a detailed view of terrain in the shortest amount of time. They can cover long distances quickly, conduct thorough inspections, and collect comprehensive data in a fraction of the time required for manual inspections.
- *Improved Safety:* Inspecting power and transmission lines is hazardous work. Using drones improves safety by minimizing risks to personnel. Drones can navigate close to power lines and infrastructure while keeping human operators at a safe distance. They provide a safer alternative to manual inspections by reducing the need for personnel to physically access hazardous or hard-to-reach areas. Manual inspections often require workers to climb transmission towers or traverse challenging terrain, putting them at risk of falls, electrical hazards, or other accidents. By deploying drones, utilities companies can minimize these risks and prioritize the safety of their personnel.
- *Improved Accuracy:* Drones equipped with high-resolution cameras, LiDAR sensors, and thermal imaging technology offer superior, accurate data collection capabilities compared to human inspectors. Drones capture detailed imagery and generate accurate 3D maps, allowing for precise identification of potential risks such as vegetation encroachment, structural deformations, or temperature anomalies.
- *Remote Accessibility:* Transmission lines are often located in remote or inaccessible areas, such as mountainous regions, dense forests, or vast expanses. Drones, with their ability to fly over diverse terrains and navigate difficult environments, provide remote accessibility to these areas.
- *Faster Inspection*: Drones inspect power lines faster. They can cover large stretches of power lines in a short amount of time. Of course, when a survey does reveal areas of concern, a drone can also slow down to take a closer look by zooming in with its cameras. Since drones perform inspections more quickly, they can also perform them more frequently. This gives you regular

reporting and increases your capacity to respond to issues and prevent outages. Drones can identify a wide range of defects and faults for a comprehensive inspection. Drone inspection data is highly accurate.
- *Data Collection:* Drones can improve the efficiency and accuracy of data collection. Drone data can also be processed faster.
- *Emergency Response*: Drones can help provide a quick response in emergency conditions.

9.5 CHALLENGES

Drones for power line inspection still face challenges such as battery autonomy and signal transmission. Climate change and extreme weather events present new challenges for electrical infrastructure. Other challenges of drones in power systems include the following [12]:

- *Regulations and Standards:* Lack of regulatory knowledge can also lead to misuse, so civil and commercial UAV operators should be trained, tested, and certified. As technology evolves, so will the regulatory framework. Continued development in drone regulations is anticipated to ensure safe operations and protect privacy and data. Collaboration between regulators, drone manufacturers and electric companies will be key to establishing safety and operational standards.
- *Hiring Pilot is Expensive*: Drone licensed pilots are highly trained and qualified to conduct transmission line inspections. It is costly to hire an experienced, licensed drone pilot. The drone pilot must have a solid understanding of the regulations governing drone operations. Knowing the local laws and regulations is essential to ensure compliance and avoid any legal issues.
- *Security:* Perhaps the deepest shadow cast by drone use for power company applications concerns their security implications. Security concerns on the cyber front are an added worry for power companies, which are already constantly grappling with the vulnerabilities of other critical components. While drones offer many beneficial inspection uses, they can also be unlawfully used for surveillance, including video, audio, and spoofing. Drones can disrupt operations if they malfunction, posing a danger to both employees and customers.
- *Safety:* This is paramount in the drone industry, driving the adoption of robust systems and redundancies to mitigate risks. Drones can be a safety risk if they crash into obstacles, encounter bad weather, or their batteries run out. Fail-safe mechanisms are integrated into drones, triggering emergency actions during signal loss or critical failures. Redundancies in critical components like flight controllers, power systems, and communication links further bolster reliability. Parachute systems can be added for controlled descents during emergencies. These comprehensive measures promote responsible and secure drone operations in diverse environments.
- *Drone Attacks*: Drones can disrupt power supplies, damage equipment, and cause safety issues by attacking power plants, electrical grids, and other energy infrastructure.
- *Start-Up Costs:* Purchasing hardware and software and training employees can be expensive, and that makes companies reluctant to commit to drones, even if the long-term savings and process improvements more than cover the start-up costs.

- *Cost-effectiveness:* Drones may not be cost-effective for some power system applications because they cannot fly far enough or long enough, or they cannot carry the necessary payloads.
- *Training:* Utility companies need to provide specialized training for drone operators and integrate drones into their existing technology systems.
- *Limitations:* Today's drones are great, but they are not perfect. For many energy and utility applications, drones cannot fly long enough or far enough or carry the necessary payloads for them to be cost-effective. For example, limited battery life restricts the time and distance a drone can fly. High expectations can make it hard for energy and utility industry leaders to understand that drones are not going to solve all of their problems.
- *Resistance to Change:* Every industry has its own methods and systems that have worked well for years and will continue to do the job. Changing these systems and adopting drones is hard, but change is essential to reap the rewards of reduced costs, better outcomes, and enhanced safety. People may resist a new solution since they have always done it a certain way.

9.6 CONCLUSION

Drones allow us to look at parts of our electric grid that are more difficult or costly to reach. Drone technology and power industry applications are growing at a rapid pace. This market segment is fast-growing with a potentially bright future. The use of drone in power systems is constantly changing, with exciting future trends and innovations on the horizon. Drones for the power industry are becoming a top-tiering technology that every industry utilizes to provide improved power line inspections. Drones that use artificial intelligence (AI) offer us many options that could make condition assessment operations even more efficient. More information about drones in power systems can be found in the books in [13-15].

REFERENCES

[1] "A bird's-eye view: Drones in the power sector," January 2018, https://www.powermag.com/a-birds-eye-view-drones-in-the-power-sector/

[2] M. N. O. Sadiku, P. A. Adekunte, and J. O. Sadiku, "Drones in power systems," *International Journal of Trend in Scientific Research and Development,* vol. 8, no. 5, September-October 2024, pp. 674-680.

[3] "The heart of a drone -- Power system," https://www.linkedin.com/pulse/heart-drone-power-system-t-drones

[4] "4 Use cases on why using drones to collect data improves inspections," December 2021, https://blog.cloudfactory.com/4-use-cases-on-why-using-drones-to-collect-data-improves-inspections

[5] "How drones work and how to fly them," May 2024, https://dronelaunchacademy.com/resources/how-do-drones-work/

[6] "What are the main applications of drones?" June 2024, https://www.jouav.com/blog/applications-of-drones.html

[7] "Drones in manufacturing: A game-changer for industry,"

[8] "Drones for inspection of power lines: Autonomy and precision," https://zmscable.es/en/drones-inspeccion-lineas-electricas/

[9] " National grid's drones: What are they used for?" https://www.nationalgrid.com/stories/grid-work-stories/national-grids-drones-what-are-they-used-for

[10] "The role of drones in enhancing transmission line maintenance," https://thedronelifenj.com/drones-transmission-line-maintenance/

[11] "5 Major benefits of drone power line inspections," June 2023, https://thedronelifenj.com/benefits-of-drone-power-line-inspections/

[12] S. Howe, "6 Barriers to drone adoption in energy & utilities and how to overcome them," January 2023, https://www.commercialuavnews.com/energy/6-barriers-to-drone-adoption-in-energy-utilities-and-how-to-overcome-them#:~:text=Systems%20Issues%3A%20Today's%20drones%20are,distance%20a%20drone%20can%20fly.

[13] A. O. Zaporozhets and Y. I. Sokol (eds.), *Control of Overhead Power Lines with Unmanned Aerial Vehicles (UAVs)*. Springer, 2021.

[14] P. V. Soni, *Characterization and Optimization of UAV Power System for Aerial and Submersible Multi-medium Multirotor Vehicle*. Rutgers University, 2016.

[15] D. Cvetković (ed.), *Drones - Various Applications*. IntechOpen, 2024.

CHAPTER 10

DRONES IN TELECOMMUNCATIONS

"The Internet is the heart of this new civilization, and telecommunications are the nervous system, or circulatory system."

- Carlos Slim

10.1 INTRODUCTION

Drones have evolved in the past few years to become exceptionally versatile across practically every industry. From improving efficiency to reducing risks, the various commercial uses for drones make them indispensable tools in many sectors. Drones can quickly identify potential issues, such as damaged equipment or vegetation encroachment, allowing for more efficient maintenance and reducing the risk of service disruptions.

Drone technology has exciting growth potential across many industries. One such industry is telecom. Telecommunications industry reaches into every corner of our economies, societies, and private lives, and it is one of the greatest drivers of economic growth and human equality. It encompasses a wide array of services and technologies, including telephone services, Internet access, and broadcasting. Telecom operators increasingly rely on drone services to inspect and maintain communication towers and other critical infrastructure. Every telco should investigate how drones can lower their cost, speed up their deliverables, reduce risk, and provide additional revenue streams.

The integration of drones with other technologies, such as AI and IoT, will further enhance their capabilities and create new opportunities for growth and optimization. Telecom operators have started drone cell tower inspections to assess network coverage and performance, map tower structures, and provide real-time data to repair crews.

Drone applications enable various types of companies to transform their operations and gain efficiency. One of the sectors that has seen a radical transformation due drone adoption is telecommunications. Drone technology is changing the face of telecommunications quickly and its first real application in telcos is tower inspections.

They are becoming increasingly popular in the telecom industry as they provide a more efficient and cost-effective way for technicians to conduct tower inspections and surveys.

The use of drones in the tower industry is growing quickly for network operators, tower engineers, tower owners, and tower climbers [1].

This chapter examines the applications of drones in the telecommunications industry. It begins with describing what a drone is. It covers telecommunications drones. It presents some applications of drones in the telecommunications industry. It highlights some benefits and challenges of drones in telecommunications. The last section concludes with comments.

Figure 10.1 A typical drone [2].

10.2 WHAT IS A DRONE?

The FAA defines drones, also known as unmanned aerial vehicles (UAVs), as any aircraft system without a flight crew onboard. Drones include flying, floating, and other devices, including unmanned aerial vehicles (UAVs), that can fly independently along set routes using an onboard computer or follow commands transmitted remotely by a pilot on the ground. A typical drone is shown in Figure 10.1 [2]. A drone is usually controlled remotely by a human pilot on the ground, as typically shown in Figure

10.2 [3]. Drones can range in size from large military drones to smaller drones. Drones, previously used for military purposes, have started to be used for civilian purposes since the 2000s. Since then, drones have continued to be used in intelligence, aerial surveillance, search and rescue, reconnaissance, and offensive missions as part of the military Internet of things (IoT). Today, drones are used for different purposes such as aerial photography, surveillance, agriculture, entertainment, healthcare, transportation, law enforcement, etc.

Figure 10.2 A drone is usually controlled by operators on the ground [3].

Commercial drones have come a long way in the last decade.

Drones work much like other modes of air transportation, such as helicopters and airplanes. When the engine is turned on, it starts up, and the propellers rotate to enable flight. The motors spin the propellers and the propellers push against the air molecules downward, which pulls the drone upwards. Once the drone is flying, it is able to move forward, back, left, and right by spinning each of the propellers at a different speed. Then, the pilot uses the remote control to direct its flight from the ground [4].

Drone laws exist to ensure a high level of safety in the skies, especially near sensitive areas like airports. They also aim to address privacy concerns that arise when camera drones fly in residential areas. These include the requirement to keep your drone within sight at all times when airborne. In the United States, drones weighing less than 250g are exempt from registration with civil aviation authorities. If your drone exceeds 250g in weight, you will also require a Flyer ID, which requires passing a test [5]. It is necessary to register as an operator, be trained as a pilot, and have civil liability insurance, in addition to complying with various flight regulations, and those of the places where their use is permitted.

Most drones have a limited payload, usually under 11 pounds. Drones are classified according to their size. Here are the different drone types:

- Nano Drone: 80-100 mm
- Micro Drone: 100-150 mm
- Small Drone: 150-250 mm
- Medium Drone: 250-400 mm
- Large Drone: 400+ mm

One of the emerging trends in drone use for factories is the utilization of LiDAR technology. LiDAR stands for Light Detection and Ranging. This technology provides accurate depth information essential for understanding the three-dimensional structure of the environment. LiDAR sensors emit laser beams to measure distances to objects, creating high-resolution 3D maps of the surrounding terrain and objects. The ability to capture detailed data through LiDAR technology has opened up opportunities for better predictive maintenance, reduction in inspection times, and overall cost savings [6].

10.3 TELECOMM DRONES

The telecommunications industry is a critical sector that facilitates the transmission of information across distances through electronic means. This industry relies on complex infrastructure such as satellites, fiber optics, and cellular networks to provide seamless communication [7]. The telecom sector accounts for 8.5 per cent of the GDP of US.

Drones have already revolutionized several industries such as business, entertainment, manufacturing, military operations, security, search and rescue, and inspections. The telecommunications industry is one of the first industries to benefit from the use of drones. Payload drones in the telecommunications industry are sometimes called cargo drones, delivery drones, transport drones, load-carrying drones, freight drones, logistics drones, utility drones, aerial transport vehicles, load-bearing UAVs (unmanned aerial vehicles), and heavy-lift drones. Telecom operators are considering whether to build their drone solutions or embrace third-party drone services for telecom drones, such as cell tower inspections and telecom tower inspections. A typical drone used for telecommunications is shown in Figure 10.3 [8].

Figure 10.3 A typical drone used for telecommunications [8].

Telecommunications inspections have become much more complex. Drone adoption is happening exponentially as the cost of drones decreases each year and the technology continues to evolve to provide autonomous flight, collision avoidance, auto landing, and communications. The uses of drone technology in the telecom sector include [9].

- Inspections of towers and tower-based equipment
- Line of sight testing
- Signal strength and coverage testing
- Maintenance and monitoring
- Leveraging drones in low-latency IoT

10.4 APPLICATIONS

Drones are increasingly utilized in the telecommunications industry. They are revolutionizing the way we monitor and safeguard our energy infrastructure. Common applications include the following [8,10].

- *Infrastructure Inspection:* Drones have become an indispensable tool for tower inspections and maintenance in the telecommunications industry. Infrastructure inspection and maintenance using commercial drones have become increasingly popular due to their ability to access hard-to-reach locations, reduce costs, and improve safety. Tower inspections is one area where drones excel and they deliver high performance inspections. It is no longer necessary for employees spend hours setting up cranes to mount a camera in the air. Drones have proven to be invaluable tools for inspecting and maintaining telecommunications infrastructure and power lines. Drones can detect structural issues or potential hazards by inspecting structures such as bridges and buildings, enabling timely and targeted repairs. Equipped with high-resolution cameras and other sensors, drones can safely fly close to transmission lines, towers, and antennas, capturing detailed images and data for analysis. Figure 10.4 shows using a drone for tower inspection [11].

Figure 10.4 Using a drone for tower inspection [11].

- *Rescue Operations:* Drones have become indispensable tools in search and rescue operations, improving the speed and efficiency with which missing persons can be located. In situations where time is of the essence, this rapid response capability can mean the difference between life and death. By reducing the time and resources required to conduct search and rescue operations, drones improve the likelihood of successful rescues.
- *Surveying and Mapping:* Drones play a critical role in surveying and mapping for the telecommunications industry. They can rapidly generate accurate 3D models and high-resolution orthomosaic maps of the terrain. Drones can cover large areas quickly, making them ideal for collecting data in remote or hard-to-reach locations.
- *Network Expansion:* As the demand for better and faster connectivity continues to grow, the telecommunications industry must constantly expand and improve its networks. Drones help achieve this by assisting in the installation of new equipment and monitoring the progress of ongoing projects. Drones also enable real-time monitoring of construction sites, ensuring that projects stay on schedule and meet quality standards.

- *Signal Strength Testing:* Drones equipped with specialized antennas and signal testing equipment can perform aerial signal strength testing. This helps engineers fine-tune the positioning and alignment of antennas, improving network performance and coverage.
- *Network Planning:* Drones play a significant role in network planning by providing accurate and up-to-date data on terrain, infrastructure, and signal strength. This information helps engineers design and optimize networks for maximum coverage and performance.
- *Emergency Response:* Drones have been utilized to improve emergency response in the telecommunications industry. In the aftermath of a natural disaster, such as a hurricane or earthquake, drones can be deployed to assess the damage to telecommunications infrastructure and identify areas where repairs are needed. This rapid assessment allows companies to restore service more quickly.

10.5 BENEFITS

The commercial use of drones is transforming various industries and providing significant benefits, from increased efficiency and cost savings to improved safety and reduced environmental impact. Drones reduce the need for manual labor and lower the risks associated with working at height or in confined spaces. A drone is unmanned, so if there is an accident in the air, only the equipment is harmed. As the benefits of drone technology in telecommunications become more widely recognized, we can expect to see greater adoption across the industry. Other benefits of using drones in telecommunications include the following [10,12,13]:

- *Connectivity:* The telecommunications industry plays a vital role in keeping us connected. As the telecom industry grows and expands, drones will be crucial in supporting its infrastructure and connectivity needs. Drones are being used to connect inaccessible areas, providing connectivity across different parts of the world. Adopting drone technology in the telecom industry is essential for developing and deploying 5G connectivity.
- *Cost Savings:* Drone technology can help reduce the costs of tower inspections and surveying. Drones offer significant cost savings in the telecommunications industry by reducing the need for manual labor and expensive equipment. They can minimize downtime by identifying issues before they cause network outages, reducing maintenance costs. By using drone technology, companies can streamline the planning process, and reduce costs,
- *Improved Safety:* Traditional methods require a team of experienced linemen to visually inspect all lines in an area to evaluate the lines for any damage or potential problems. Utilities are increasingly turning to drones for the inspection of powerlines and pipelines in hard-to-reach areas or in difficult terrain. Using drones in telecommunications greatly enhances worker safety. Climbing towers for inspections and maintenance can be dangerous, and accidents can result in injuries or even fatalities. Drones eliminate the need for technicians to climb towers in many cases, as typically shown in Figure 10.5 [14]. This reduces the risk of accidents and promoting a safer work environment. Drones equipped with cameras and sensors can provide additional security for telecom infrastructure, helping to detect and deter potential vandalism or theft.

Figure 10.5 Drones eliminate the need for technicians to climb towers [14].

- *Increased Efficiency:* Accurate cell tower inspections in a matter of minutes using drones. Drones can perform tasks more quickly and accurately than humans, increasing efficiency in the telecommunications industry. They can inspect towers, map terrain, and test signal strength in a fraction of the time it would take a team of technicians.
- *Better Data Collection:* Companies large and small seek ways to collect data faster and lower risk and cost. Drones equipped with advanced sensors and cameras can collect valuable data that can be used to optimize networks, monitor infrastructure, and plan future expansions. The data collected provides companies with the information they need to make data-driven decisions and improve their operations.
- *Real-time Data:* Drones can provide real-time data and analytics and help decrease tower downtime. Real-time aerial data analytics for tower inspection allows operators to collect, process, and analyze data in real-time. Drones for tower inspection can capture high-resolution images and data from difficult-to-reach areas and use real-time data analytics and artificial intelligence (AI) to analyze structural data and identify problem areas. Telcos can use real-time data from drones to improve wireless network performance, identifying and addressing problem areas, and quickly resolving issues to improve the overall quality of service.

- *Access to Remote Areas*: Drones can reach areas that are difficult to access, allowing telecom providers to expand their geographic presence. They can quickly access hard-to-reach areas, such as cell towers or transmission lines, facilitating timely repairs and minimizing the need for manual inspections.
- *Faster Inspection:* Drones have become an indispensable tool for tower inspections and maintenance in the telecommunications industry. They can be used to conduct inspections more frequently, allowing for early detection of small issues. They can cover a larger area in a shorter amount of time than traditional inspection methods. Drones can gather information more efficiently than human inspectors, and the digital data can help employees make better decisions.
- *Reduced Risk:* Drones can speed up message and response times, which can reduce the risk of damage to pipelines. They significantly improve safety, save time, and minimize downtime.
- *Surveying:* Drones play a critical role in surveying and mapping for the telecommunications industry. They can rapidly generate accurate 3D models and high-resolution orthomosaic maps of the terrain, helping engineers identify the best locations for new towers or network expansions. Drones can cover large areas quickly, making them ideal for collecting data in remote or hard-to-reach locations. This accelerates the planning process and ensures optimal network coverage.
- *Network Expansion:* As the demand for better and faster connectivity continues to grow, the telecommunications industry must constantly expand and improve its networks. Drones help achieve this by assisting in the installation of new equipment and monitoring the progress of ongoing projects. They can transport small payloads, such as antennas and cables, to hard-to-reach locations, speeding up the deployment process.
- *Timely Maintenance:* Drones are used for timely maintenance and repairs in the telecommunications industry. Drones equipped with remote monitoring capabilities allow for efficient and accurate inspections of telecommunication infrastructure. By employing predictive analytics, drones can identify potential issues before they lead to costly downtime or service interruptions. Through real-time data collection and analysis, drones can detect anomalies in equipment performance or identify areas requiring maintenance.
- *Environmental Benefits*: By incorporating drones into operations, the telecommunications industry can reap various environmental benefits. Drones have the potential to revolutionize wildlife monitoring and precision agriculture, leading to a more sustainable future. By identifying areas of crop stress, drones can enable targeted interventions, reducing the overall environmental impact of agriculture. Incorporating drones into operations can thus contribute to the preservation of biodiversity and the promotion of sustainable farming practices.
- *Competitive Advantage*: Incorporating drones into operations in the telecommunications industry can provide a significant competitive advantage. Drones have the potential to revolutionize the way businesses in this sector operate, offering strategic differentiation and the opportunity for market dominance. Embracing drone technology positions businesses in the telecommunications industry as innovative and forward-thinking, helping them gain a competitive edge in the market.

10.6 CHALLENGES

There are several challenges facing the use of drones in telecommunications. The main challenge is the need for skilled drone operators with experience in the telecom and drone industries. Access to high ground in rural areas is another challenge. Drones will not replace engineers nor tower climbers, but they are enabling them to work faster and more efficiently. Drone technology also comes with privacy concerns, regulatory challenges, and potential security risks. Other challenges facing drones in telecommunications include the following [10,15]:

- *Privacy:* Privacy is perhaps the most controversial matter when it comes to malicious drone use. Drones used in the telecommunications industry often collect large amounts of data, including images and videos of infrastructure and terrain. Operators must comply with privacy regulations and ensure that any collected data is stored and used in accordance with applicable laws. Drone operators can help ensure the safe and responsible integration of drones into the telecommunications industry.
- *Regulation:* As drones become more common in the telecommunications industry, it is essential to be aware of the regulations that govern their use. These regulations are in place to ensure the safety of both drone operators and the general public. Since drone usage is still a relatively new, several regulatory issues and grey areas must be addressed to integrate drones into the telecom industry successfully. As drones become more common in the telecommunications industry, it is essential to be aware of the regulations that govern their use. These regulations are in place to ensure the safety of both drone operators and the general public. Drones used for commercial purposes, such as in the telecommunications industry, must be registered with the relevant aviation authority. As drones become more integrated into the telecommunications industry, regulations will need to adapt to accommodate these new applications.
- *Airspace Restrictions:* Drone operations in the telecommunications industry must adhere to airspace restrictions and follow any applicable rules for flying near airports, heliports, or other sensitive areas.
- *Operator Licensing:* Telecom operators can carry out drone flights to execute necessary tasks, such as collecting images and video footage. Drone operation requires skilled pilots and regulatory compliance, which can be challenging for telecom operators. In the US, Canada, and other parts of the world, all drone operators must be duly licensed and maintain their piloting education to satisfy the rigorous criteria. Drones must adhere to specific aviation and privacy regulations.
- *Hatred for Drones:* Drone pilots are probably familiar with the experience of drawing small crowds whenever they are out flying drones. However, the opposite of this experience is also not uncommon, people demanding that these drones stay away from them. This is a story that gets told over and over, a person shoots down a drone that is flying near or above their property because of concerns over their safety or privacy, or simply because they believe that the drone should not be flying over their private property.
- *Drones for Evil Means*: While drones have proven essential for business and humanitarian operations. it is not impossible for drones to be used for evil means. Just like any technology, drones can be either good or bad depending on how people use them.
- *Expensive Technology*: As with any technology that pushes the boundary of what is possible, many of these advanced drones remain firmly beyond the budget of most people or organizations.

Buying advanced equipment can provide a competitive edge, but these are certainly too expensive for a drone pilot who is still early in their career. As the commercial drone industry becomes more competitive, it becomes apparent that a drone pilot needs to have a somewhat expensive drone to even be noticed by clients. Add to that the costs of certification and training and you are really going to need significant investment to even get started in a professional drone career.

- *Skill Shortage*: There may be a lack of skilled drone engineers to set up and operate drone systems.
- *Communication:* Drones can face challenges with communication such as channel conditions, interference, antenna orientation, intermodulation distortion, spectrum availability, and network congestion. For example, when multiple drones and terminals share a wireless network, it can cause congestion.
- *Signal Interference:* Signal interference can limit the effectiveness of drones.
- *Logistics:* Drone delivery could pose logistical issues, especially in densely populated areas. The technology would need to be capable of maneuvering around buildings and wildlife like birds. Also, deliveries to specific apartments would be difficult without human intervention. So you can get almost anywhere fairly quickly by truck or car. That makes it much harder for drones because the drones now have to compete against something that is been fine-tuned and optimized.
- *Vulnerability to Hacking*: Hackers can attack a drone's central control system and gain control of the drone.

10.7 CONCLUSION

Telecom operators should develop a tailored strategy and implementation road map for commercial drone applications. As drones in telecommunications industry become increasingly prevalent, they are transforming the way companies approach tower inspections, network expansions, and emergency response. As drone technology continues to advance, we can expect greater integration with other technologies, such as artificial intelligence (AI) and the Internet of things (IoT). The widespread deployment of 5G bandwidth will also improve drone performance. The bandwidth of 5G provides drones the ability to upload cloud imaging in real time. We can also expect to see even more innovative applications in the telecommunications industry.

The future of telecommunications is definitely wireless. The future of telecom drones looks bright and promising as drone technology continues to evolve and improve. Future advancements in drone technology could lead to improved capabilities and new applications in the telecommunications industry.

REFERENCES

[1] M. N. O. Sadiku, O. D. Olaleye, and J. O. Sadiku, "Drones in telecommunications," *International Journal of Trend in Research and Development,* vol. 11, no. 5, September 2024, pp. 102-105.

[2] D. Daly, "Tethered drones: A critical component of tactical communications," September 2021,

[3] "Revolutionize your inspections with WISPR American-made drones," https://wisprsystems.com/drones-for-inspection/?gad_source=1&gclid=EAIaIQobChMI9LvR9KyniAMV4ZHCCB3JSygnEAAYASAAEgKFyPD_BwE

[4] "How drones work and how to fly them," May 2024, https://dronelaunchacademy.com/resources/how-do-drones-work/

[5] "What are the main applications of drones?" June 2024, https://www.jouav.com/blog/applications-of-drones.html

[6] "Drones in manufacturing: A game-changer for industry," https://viper-drones.com/industries/infrastructure-drone-use/manufacturing/#:~:text=The%20integration%20of%20drones%20into,on%20manufacturing%20is%20no%20exception.

[7] "Applications of payload drones in the telecommunications industry," https://gaotek.com/applications-of-payload-drones-in-telecommunications-industry/

[8] "Top 10 commercial uses for drones," April 2023, https://www.inspiredflight.com/news/top-10-commercial-uses-for-drones.php

[9] S. D. Sree, "Drone technology in the telecom sector," January 2023, https://dronefluence.com/drone-technology-in-the-telecom-sector/

[10] C. Guarnera, "Connecting the world: How drones are revolutionizing the telecommunications industry," May 2023, https://www.bluefalconaerial.com/drones-in-telecommunications-industry/

[11] "The bright & growing future of drones in the telecom industry," https://pointivo.com/future-of-drones-in-the-telecom-industry/

[12] C. Guarnera, "Connecting the world: How drones are revolutionizing the telecommunications industry," May 2023, https://www.bluefalconaerial.com/drones-in-telecommunications-industry/

[13] "Advantages of using drones in the telecommunications industry," https://www.av8prep.com/aviation-library/part-107-drone/advantages-of-using-drones-in-the-telecommunications-industry

[14] "3 Reasons you should use drones for cell tower inspections," https://consortiq.com/uas-resources/3-reasons-you-should-use-drones-for-cell-tower-inspections

[15] "The pros and cons of drone technology," https://pilotinstitute.com/drone-pros-and-cons/

CHAPTER 11

DRONES IN LAW ENFORCEMENT

"The police are the public and the public are the police; the police being only members of the public who are paid to give full-time attention to duties which are incumbent on every citizen in the interests of community welfare and existence."

—Robert Peel

11.1 INTRODUCTION

Drones, or unmanned aircraft systems, are a fusion of technology, engineering, and innovation. These unmanned aerial vehicles (UAVs) have transcended their initial roles, emerging as pivotal tools in law enforcement. From enhancing aerial photography to revolutionizing agriculture, drones have now found a significant role in policing. Across the United States, police departments are increasingly harnessing the power of drones [1]. In the United States, over 1,000 police departments already use drones.

In any major city you will find drones supporting the work of police departments, fire departments, search and rescue teams, and disaster management personnel. While some drones are good for their autonomy and suited for quickly searching an area for victims after a crash, other drones are good for collecting LiDAR data in rugged environments to support the creation of detailed 3D maps for disaster planning or risk mitigation. Drones can provide a real-time aerial view of evacuation routes and help in monitoring the movement of people and vehicles. They can quickly create detailed maps of affected areas, providing emergency responders with essential information for planning and coordinating their efforts.

As drone technology becomes more prevalent and specialized, drones and supporting software are developed specifically for the needs of several types of jobs, including law enforcement. Drones provide law enforcement with a bird's-eye view of unfolding situations in real time, allowing officers to gather critical information without putting themselves in harm's way. They are making incident response operations not only safe and effective but also remotely visible and manageable.

The use of drones by law enforcement agencies has grown exponentially in the past few years. This is partly due to advances in drone technology and also due to the fact that drones are becoming more affordable and accessible. Thus, drones are proving to be an essential tool for law enforcement agencies, enabling them to respond to emergencies safely, quickly, and efficiently. From search and rescue to crime scene investigation, the use cases for police drones run the gamut of law enforcement missions [2].

This chapter highlights the applications of drones in law enforcement. It begins with describing what a drone is. It cover police drones. It presents some applications of drones in law enforcement. It highlights some benefits and challenges of drones in law enforcement. The last section concludes with comments.

Figure 11.1 A typical drone [3].

11.2 WHAT IS A DRONE?

The FAA defines drones, also known as unmanned aerial vehicles (UAVs), as any aircraft system without a flight crew onboard. Drones include flying, floating, and other devices, including unmanned aerial vehicles (UAVs), that can fly independently along set routes using an onboard computer or follow commands transmitted remotely by a pilot on the ground. A typical drone is shown in Figure 11.1 [3].

A drone is usually controlled remotely by a human pilot on the ground, as typically shown in Figure 11.2 [4]. Drones can range in size from large military drones to smaller drones. Drones, previously used for military purposes, have started to be used for civilian purposes since the 2000s. Since then, drones have continued to be used in intelligence, aerial surveillance, search and rescue, reconnaissance, and offensive missions as part of the military Internet of things (IoT). Today, drones are used for different purposes such as aerial photography, surveillance, agriculture, entertainment, healthcare, transportation, law enforcement, etc.

Figure 11.2 A drone is usually controlled by operators on the ground [4].

Commercial drones have come a long way in the last decade. Drones work much like other modes of air transportation, such as helicopters and airplanes. When the engine is turned on, it starts up, and the propellers rotate to enable flight. The motors spin the propellers and the propellers push against the air molecules downward, which pulls the drone upwards. Once the drone is flying, it is able to move forward, back, left, and right by spinning each of the propellers at a different speed. Then, the pilot uses the remote control to direct its flight from the ground [5].

Drone laws exist to ensure a high level of safety in the skies, especially near sensitive areas like airports. They also aim to address privacy concerns that arise when camera drones fly in residential areas. These include the requirement to keep your drone within sight at all times when airborne. In the United States, drones weighing less than 250g are exempt from registration with civil aviation authorities. If your drone exceeds 250g in weight, you will also require a Flyer ID, which requires passing a test [6]. It is necessary to register as an operator, be trained as a pilot, and have civil liability insurance, in addition to complying with various flight regulations, and those of the places where their use is permitted.

Most drones have a limited payload, usually under 11 pounds. Drones are classified according to their size. Here are the different drone types:

- Nano Drone: 80-100 mm
- Micro Drone: 100-150 mm
- Small Drone: 150-250 mm
- Medium Drone: 250-400 mm
- Large Drone: 400+ mm

One of the emerging trends in drone use for factories is the utilization of LiDAR technology. LiDAR stands for Light Detection and Ranging. This technology provides accurate depth information essential for understanding the three-dimensional structure of the environment. LiDAR sensors emit laser beams to measure distances to objects, creating high-resolution 3D maps of the surrounding terrain and objects. The ability to capture detailed data through LiDAR technology has opened up opportunities for better predictive maintenance, reduction in inspection times, and overall cost savings [6].

11.3 POLICE DRONES

Police drones are a new form of modern policing and play an important role in supporting law enforcement and maintaining public safety. Police law enforcement drones can be used for surveillance and reconnaissance, search and rescue, traffic supervision, incident scene investigation, emergency response, etc. Police drones are quickly becoming an essential tool for law enforcement agencies around the world because help provide reliable performance in diverse and challenging environments.

A police drone, also known as cop drones or law enforcement drones, is any drone used by law enforcement. The drone is specifically made for use by the police. It has become indispensable tools in modern policing. The use of drones in law enforcement began way back in the early 2000s. However, it was not until the mid-2010s that drone technology became widely accessible and affordable for police departments. Since then, more and more police departments have started using drones. Law enforcement has also borrowed drones used for the military. There are so many uses for police drones that they can vary widely in size and capability. Police drones programs become a prominent trend in law enforcement, showcasing the evolving capabilities and widespread acceptance of drone technology in public safety [8]. Using police drones has caused significant improvements in response times and situational awareness. A police drone in operation is shown in Figure 11.3 [9]. Popular cop drones include BRINC LEMUR 2, Skydio X2, DJI Matrice 350 RTK, Parrot ANAFI USA, and DJI Mavic 3 Thermal. A variety of police drones are displayed in Figure 11.4 [9].

Figure 11.3 A police drone in operation [9].

Figure 11.4 A variety of police drones [9].

Police drones are designed to withstand harsh environments and adverse weather conditions. There are some characteristics most police drones share [7]:

- *Advanced sensors/cameras* for high-quality visual, thermal, and other data collection
- *Communication systems* that enable real-time data collection and transmission
- *Rugged design* made for operations in harsh environments

The primary functions of drones in policing include [8]:

- Automatic response to 911 calls
- Surveillance and reconnaissance
- Search and rescue
- Traffic accident reconstruction
- Traffic monitoring
- Crime scene investigation
- Crowd control and monitoring
- Tactical operations
- Drug and smuggling interdiction
- Evidence gathering
- Disaster response and damage assessment

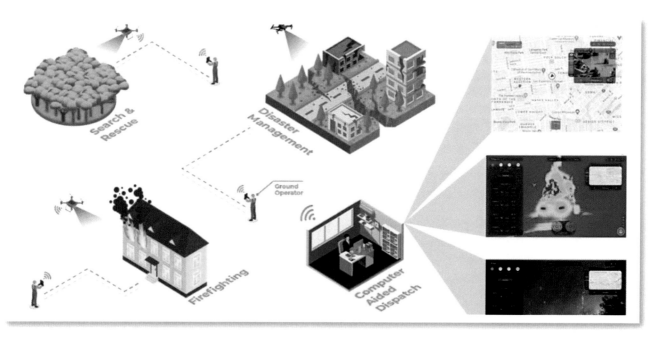

Figure 11.5 Some uses of drones by law enforcement [9].

11.4 APPLICATIONS

We now look at some real-world examples of how police departments and other law enforcement agencies are actually using drones in their work. Drones have been used here in several ways by police departments and law enforcement agencies. Some of the uses are shown in Figure 11.5 [9]. Common applications of drones in law enforcement include the following [4,9]:

- *Drone as First Responder:* Perhaps the most promising deployment of police drones lies in the drone as a first responder (DFR) program. Traditionally, first responders such as police, firefighters, and emergency medical teams are dispatched to the scene of an incident as soon as a distress call is received. In a DFR program, a drone is deployed to the scene immediately after a 911 call is received, often arriving before human responders. The police force is seamlessly implementing drone as a DFR program and revolutionizing the emergency response strategy. The DFR program launches a drone at the time of 911 dispatch and is designed to provide eyes on the scene prior to the arrival of ground units. The drone provides invaluable real-time situational awareness by streaming live video to officers before they reach the scene. The DFR program not only improves the efficiency and effectiveness of emergency response operations but also enhances the safety of the first responders by providing them with critical information before they arrive on the scene.
- *Search and Rescue Operations:* The use of drones in search and rescue operations is particularly notable. A thermal drone is commonly used by police to search for missing persons and animals. Drones can quickly acquire aerial data over large areas and pinpoint where missing people may be trapped. Drones can get to a location faster than ground vehicles and reach inaccessible areas.
- *Accident Investigations*: Drones have become an important tool for accident scene investigation. An advantage of using drones in accident scene analyzing is the level of detail and accuracy they provide. 3D reconstructions from drone pictures provide a detailed view of a scene. Drones in forensic investigations of incident scenes provide great value by capturing information from a top-down view and data from scenes that may have been missed from the ground. Deploying drones allows law enforcement to collect more informational data in real-time.
- *Traffic Management:* In an era where technology continually reshapes the landscape of public safety, drones have soared into the spotlight, especially in traffic management. Drones provide law enforcement agencies with a versatile tool for comprehensive traffic monitoring, enhancing response capabilities and ensuring safer roads for everyone. Police drones can also be used for traffic management in rush hour and congested areas. It may be difficult for police officers on the ground to assess the causes and dynamics of traffic congestion, but a drone can easily solve this problem by flying overhead to determine the cause of traffic jams.
- *Crime Analysis:* Flying around crime scenes, drones can collect photos at different heights and angles. Compared to traditional human forensics, the speed with which drones can move from place to place and collect an uninterrupted stream of data is unparalleled. In today's active time of shootings and terrorist attacks, police forces are taking advantage of drones to safely and quickly arrest suspects.
- *Crowd Monitoring:* Replacing or complementing traditional helicopter surveillance, drones present a more cost-effective and less intrusive method of aerial monitoring. Drone surveillance provides a broader view of crowd gatherings and events and transmits real-time data to crowd control teams. Drones can zoom in on areas of interest and provide officers with detailed information about what is happening on the ground. This information can help make critical decisions for the control team. A typical example of crowd monitoring is shown in Figure 11.6 [4].

Figure 11.6 A typical example of crowd monitoring [4].

- *Tethered Drones:* Tethered drones are another tool being deployed by law enforcement and other public safety agencies. They offer distinct advantages in event security. The drones are deployed in a stationary mode, which allows continuous overwatch. The tether serves as its power source and negates the need to change batteries. Tethered drones can be mounted on a vehicle, in a compartment, or be portable in a Pelican case and launched in seconds.
- *Damage Assessment:* After a natural disaster such as a hurricane, earthquake, or flood, drones can quickly assess the damage to infrastructure, buildings, and roads. They can provide real-time images and videos that help in understanding the scale of the damage and prioritizing response efforts. After a disaster, finding and rescuing survivors is the top priority. Drones equipped with thermal cameras can locate survivors in hard-to-reach areas, at night, or under debris. Figure 11.7 shows a drone for assessing damage after a disaster [9].

Figure 11.7 A drone assessing damage after a disaster [9].

11.5 BENEFITS

The capability of the drone technology has helped in various scenarios, including locating suspects, monitoring large events, assessing crime scenes, and significantly reducing the number of calls officers respond to. Drones programs have been particularly effective in urban areas, where traffic congestion and complex environments can delay traditional response times. By deploying drones to incident scenes, police departments can reduce the need for multiple ground units and allocate resources more strategically. Drones can cover a large area easily, and they can be equipped with thermal sensors. These make them effective in search and rescue operations.

Other benefits of using drones in law enforcement include [4,10]:

- *Rapid Deployment:* Drones are small and compact, and these design features allow them to be deployed quickly. Law enforcement officers can have them in the air in minutes and allow them to monitor and capture video and images
- *Cost-effectiveness:* Drones are a cost-effective way to conduct aerial surveillance. The conventional method of deploying a manned helicopter is expensive and time-consuming, and may not be suitable for situations that require an immediate response. By contrast, a drone does not require paying staff to control and monitor it, nor does it require high maintenance costs.

- *Reduced Risk:* Drones can enter a dangerous environment and provide decision-makers with a real-time view of what is happening. The appropriate team members are then dispatched afterward, helping officers avoid dangerous situations.
- *Improving Efficiency:* Police drones can cover a large area in a short amount of time. They deliver useful information faster than officers on the ground. This allows for helping officers to respond to incidents more effectively and faster than conventional vehicles.
- *Drones Save Lives:* Drones can be equipped with various attachments depending on the task. They can save lives.
- *Real-time Visuals*: In the era of information and real-time action, drones have become a significant boon to law enforcement agencies. Equipped with high-resolution cameras, these drones offer a bird's-eye view of situations as they unfold. By providing clear, live visuals, officers can make informed decisions, ensuring timely interventions.
- *Evidence Collection:* Drones can quickly capture hundreds of photos from multiple angles, which can help investigators clear a crime scene and gather evidence. Drones can also create 3D models of crime scenes for later analysis.
- *Monitoring:* In the fight against environmental crimes, drones are invaluable. Large public gatherings, like concerts, protests, or festivals, pose unique challenges to law enforcement. Drones offer a high vantage point, providing aerial views of these gatherings. With this bird's-eye view, officers can monitor crowd movements, detect potential disturbances, and deploy resources efficiently, ensuring public safety with efficiency. Drones can monitor large crowds or areas for suspicious activity. They can access hard-to-reach locations more quickly than people.
- *Border Surveillance:* Borders are sensitive areas, often requiring round-the-clock monitoring. Drones, with their extended flight times and advanced cameras, have become essential tools in border surveillance. They help detect and prevent illegal activities and crossings, ensuring a nation's security without overburdening human resources.
- *Night Vision:* The cover of night has often posed challenges to policing, with visibility dropping and operations becoming more challenging. Drones with night vision can capture images based on heat signatures which is particularly crucial for nighttime operations or in areas with dense foliage where traditional vision might falter. This capability ultimately increases situational awareness of officers and their safety. An example of drone operating at night is shown in Figure 11.8 [11].

Figure 11.8 A drone operating at night [11].

11.6 CHALLENGES

Drone use is a complicated issue, bringing with it privacy and safety concerns. Having solid policies and procedures in place to guide law enforcement drone use is key to ensuring their legal, safe use. Drones have limited range, can be expensive, and police departments must follow extensive Federal Aviation Administration (FAA) regulations around flights. The public perception of police drones raises fear. Other challenges of using drones in law enforcement include the following [12,13]:

- *Privacy Rights:* Probably the most significant issue associated with the use of drones (at least for law enforcement) is privacy. When most people hear the words "drone" and "government" together, they immediately think of the predator drone used in theaters of combat: a weaponized UAV that can fly thousands of feet in the air surveilling the public undetected. This misconception leads to fear of the drone technology. As with other technologies, addressing privacy concerns surrounding drones involves a balance of policy and engagement. Drone operators should adhere to FAA guidelines and avoid intentionally recording or transmitting images of any location where a person would have a reasonable expectation of privacy.
- *Regulation:* Police drone use is increasing, making it essential for agencies to adopt a sound law enforcement drone policy governing their use. The use of weapons on drones by police is generally not allowed due to legal and safety reasons. In addition, many countries have strict laws and regulations regarding the use of force by law enforcement, and the use of weaponized drones could be in violation of these laws. Any licensed drone can fly at that height over any property. Still, the intricacies of the law will defer from nation to nation.

- *Weaponization:* US Federal legislation prohibits weaponized drones, but several states have taken this even further. Some states also bar the use of armored drones.
- *Complexity:* Today, drones are more complex, playing host to cameras, GPS devices, infrared sensors, and more. A modern, human-operated drone can take pictures, perform search-and-rescue operations, and even serve as a first responder in areas too remote for land vehicles, and new applications are being developed every day. A drone may require a licensed pilot to fly, adding to the cost of its operation. Law enforcement agencies must ensure that their officers are properly trained and are complying with federal and state guidelines.
- *Restrictions:* For obvious reasons, several states have restrictions regarding drone operations over or near jails and prisons. This is a continuing problem, as drones have been used to deliver drugs, weapons, and other contraband to inmates in corrections facilities. Some states ban drones with facial recognition capabilities, with exceptions for search and rescue and disaster-related services.
- *Data Retention:* Data retention issues are challenging when it comes to drone use. Will all video from the drone be recorded and if so, where will it be retained and for how long? Can your agency freely share or disclose information gathered by the drone with other governmental agencies? Some states have issued specific laws that requires law enforcement agencies to destroy all information gathered by a drone within 30 days.
- *Limited Flight Time*: A major roadblock for the integration of drones into policing is the limited flight time of drones due to current battery technology. The average battery life of most drones sustains around 30 minutes of flight time. UPS, Chevron, Amazon, and other companies are looking to combat this limited flight time to make autonomous UAVs effective.
- *Public Perception:* Drones can have a negative connotation and raise fears in the public. They can injure people or property, and violate privacy rights.
- *Legal Issues*: Due to a sharp rise in the use of drones, there exists several potential legal issues connected with their use. Police departments must follow extensive FAA regulations around drone flights. The intentional filming or recording of any data could lead to serious legal problems for the operator depending on the manner of use and the ultimate use of the information gathered, unless all laws, rules, and regulations are strictly followed.
- *Pilot's Responsibilities:* The operator of the drone, including a law enforcement officer, is responsible to comply with all applicable federal and state laws, rules and regulations when operating the UAV.
- *Limited Range*: Drones have limited range and battery life, which can restrict how far they can travel and how long they can fly.
- *Threats from Criminals:* Drones can be used for more than their original purpose. Criminals and terrorists can exploit drones for illicit surveillance, chemical, biological, and radiological attack.

11.7 CONCLUSION

Drones are rapidly being adopted by police departments and law enforcement agencies all over the world, making their work significantly efficient, safer, and easier. As drones continue to evolve, law enforcement agencies around the world are using them in more ways to save lives and enhance the safety of officers. The use of drones as surveillance tools and first responders is a fundamental shift in policing. Drones have become a valuable asset for police departments, allowing them to perform tasks at a lower cost and assist in surveillance and investigations.

Overall, using drones in law enforcement helps improve public safety and operational efficiency by opening up new possibilities. Drone technology is improving and will help law enforcement more. As the technology gets better, drones will be the future of law enforcement. More information about drones in law enforcement can be found in the books in [14-20].

REFERENCES

[1] "How drones help law enforcement," https://enterprise-insights.dji.com/blog/drones-help-law-enforcement#:~:text=They%20provide%20real%2Dtime%20aerial,the%20use%20of%20manned%20aircraft.

[2] M. N. O. Sadiku, P. O. Adebo, and J. O. Sadiku, "Drones in law enforcement," *International Journal of Trend in Research and Development*, vol. 11, no. 5, September-October 2024, pp. 71-78.

[3] "What is a public safety drone?" https://uavcoach.com/public-safety-drone/#:~:text=Public%20safety%20drones%20are%20any,the%20work%20of%20first%20responders.

[4] "Top police drones | Best drones for law enforcement," June 2024, https://www.jouav.com/blog/police-drone.html

[5] "How drones work and how to fly them," May 2024, https://dronelaunchacademy.com/resources/how-do-drones-work/

[6] "What are the main applications of drones?" June 2024, https://www.jouav.com/blog/applications-of-drones.html

[7] "Drones in manufacturing: A game-changer for industry," https://viper-drones.com/industries/infrastructure-drone-use/manufacturing/#:~:text=The%20integration%20of%20drones%20into,on%20manufacturing%20is%20no%20exception.

[8] Z. Dukowitz, "Police drones: A guide to how law enforcement uses drones in its work," May 2024, https://uavcoach.com/police-drones/

[9] "Drones for law enforcement: Benefits and use cases," https://www.flytbase.com/blog/drones-for-law-enforcement

[10] "Benefits of drones in policing," October 2023, https://www.flymotionus.com/posts/benefits-of-drones-in-policing-an-in-depth-exploration

[11] "Team drone challenges | Drone entertainment," https://tlciscreative.com/new-experiences/aerial-drones-special-events/

[12] "Key considerations for a law enforcement drone policy," https://www.police1.com/police-products/police-drones/key-considerations-for-a-law-enforcement-drone-policy

[13] "Potential legal problems with drones,"
[14] https://smithwelchlaw.com/potential-legal-problems-with-drones/
[15] S. Anderson, *Privacy by Design: An Assessment of Law Enforcement Drones*. Georgetown University, 2014.

[16] D. R. Faust, *Drones for the Police and Military.* Rosen Publishing Group, 2019.

[17] A. C. Cunningham, *Drones, Surveillance, and Targeted Killings.* Greenhaven Publishing, 2017.

[18] C. Schloe, *Drones - UAS for Emergency Response Services: UAV Guide for First Responders in Law Enforcement, Park Rangers, Fire & Rescue, Emergency Management and Search & Rescue.* Christopher J. Schloe, 2017.

[19] Committee on the Judiciary United States Senate, *The Future of Drones in America: Law Enforcement and Privacy Considerations.* CreateSpace Independent Publishing Platform, 2014.

[20] D. R. Faust, *Police Drones.* Rosen Publishing Group, 2015.

[21] L. L. Bella, *Drones and Law Enforcement.* Rosen Publishing Group, 2106.

CHAPTER 12

DRONES IN ENTERTAINMENT

"The world is a stage, the stage is a world of entertainment."

- Howard Dietz

12.1 INTRODUCTION

Since the establishment of the film industry, storytellers, screenwriters, and producers have always looked for ways to make movies all the more enjoyable. The film and entertainment industry has always been at the forefront of technological innovation, constantly seeking new ways to captivate audiences and deliver breathtaking visuals.

Drone technology has proven helpful in many sectors such as education, business, agriculture, aerospace, law enforcement, construction, mining, telecommunications, and the military. In recent years, drones have emerged as powerful tools in entertainment industry, revolutionizing the way movies, TV shows, music videos, and live events are shot. Drones are now changing the way movie makers operate. The drone technology is literally changing the way movie makers operate and changing how Hollywood, the Mecca of movie-making, produces films for public consumption. It has opened up a world of creative possibilities for storytelling, allowing filmmakers to craft compelling narratives that were once limited by traditional filming methods. Drones bring a new dimension to the viewing experience by offering dynamic shots that traditional cameras cannot achieve [1].

The comprehensive features of modern drones make them highly prominent in advancing the world. Filmmakers now have a powerful tool at their disposal to capture stunning visuals from unique and dynamic perspectives. Drones are used to amplify the concert experience. They bring a new dimension to the viewing experience by offering dynamic shots that traditional cameras cannot achieve. The media and entertainment industry has been quickly rising as a result of the increased use of drones in cinematography and photography. The popularity of drones in the media and entertainment industry has skyrocketed, especially for movie production, news coverage, and documentary filming. Drone have also quickly become an integral aspect of professional photography and filming [2].

This chapter highlights the applications, benefits, and challenges of drones in entertainment. It begins with describing what a drone is. It covers entertainment drones. It presents some applications of drones in entertainment. It highlights some benefits and challenges of drones in entertainment. The last section concludes with comments.

12.2 WHAT IS A DRONE?

The FAA defines drones, also known as unmanned aerial vehicles (UAVs), as any aircraft system without a flight crew onboard. Drones include flying, floating, and other devices, including unmanned aerial vehicles (UAVs), that can fly independently along set routes using an onboard computer or follow commands transmitted remotely by a pilot on the ground. A typical drone is shown in Figure 12.1 [3]. A drone is usually controlled remotely by a human pilot on the ground, as typically shown in Figure 12.2 [4]. Drones can range in size from large military drones to smaller drones. Drones, previously used for military purposes, have started to be used for civilian purposes since the 2000s. Since then, drones have continued to be used in intelligence, aerial surveillance, search and rescue, reconnaissance, and offensive missions as part of the military Internet of things (IoT). Today, drones are used for different purposes such as aerial photography, surveillance, agriculture, entertainment, healthcare, transportation, law enforcement, etc.

Figure 12.1 A typical drone [3].

Figure 12.2 A drone is usually controlled by operators on the ground [4].

Commercial drones have come a long way in the last decade. Drones work much like other modes of air transportation, such as helicopters and airplanes. When the engine is turned on, it starts up, and the propellers rotate to enable flight. The motors spin the propellers and the propellers push against the air molecules downward, which pulls the drone upwards. Once the drone is flying, it is able to move forward, back, left, and right by spinning each of the propellers at a different speed. Then, the pilot uses the remote control to direct its flight from the ground [5].

Drone laws exist to ensure a high level of safety in the skies, especially near sensitive areas like airports. They also aim to address privacy concerns that arise when camera drones fly in residential areas. These include the requirement to keep your drone within sight at all times when airborne. In the United States, drones weighing less than 250g are exempt from registration with civil aviation authorities. If your drone exceeds 250g in weight, you will also require a Flyer ID, which requires passing a test [6]. It is necessary to register as an operator, be trained as a pilot, and have civil liability insurance, in addition to complying with various flight regulations, and those of the places where their use is permitted.

Most drones have a limited payload, usually under 11 pounds. Drones are classified according to their size. Here are the different drone types:

- Nano Drone: 80-100 mm
- Micro Drone: 100-150 mm
- Small Drone: 150-250 mm
- Medium Drone: 250-400 mm

- Large Drone: 400+ mm

One of the emerging trends in drone use for factories is the utilization of LiDAR technology. LiDAR stands for Light Detection and Ranging. This technology provides accurate depth information essential for understanding the three-dimensional structure of the environment. LiDAR sensors emit laser beams to measure distances to objects, creating high-resolution 3D maps of the surrounding terrain and objects. The ability to capture detailed data through LiDAR technology has opened up opportunities for better predictive maintenance, reduction in inspection times, and overall cost savings [7].

12.3 ENTERTAINMENT DRONES

Everybody likes entertainment. Children are waiting for fireworks, adults go to art performances, young people are always happy to see popular artists and everyone is always waiting for something new that will inspire them with ideas. Each time the drone show evokes a number of enthusiastic emotions and each time there will be a viewer who sees it for the first time.

The entertainment industry has often pioneered developments in technology. Drones are the latest technology in the entertainment industry. Entertainment drones are used in film making, at theme parks, sporting events, and more. Drone enthusiasts who want to use drones to capture beautiful moments and have fun with family and friends can purchase these recreational drones without breaking the bank.

There are different kinds of entertainment drones that can record photos or videos in HD. They can be equipped with different features, GPS, people tracking, route programming, etc. Racing or competitive drone is generally used on private and strictly regulated land. Ready-to-use drones that are available for users to assemble themselves. GPS drones are connected to satellites and use them as a means of plotting flight direction. Professional drones can be equipped with cameras or video cameras for all types of aerial photography. They are the only aircraft that can fly over built-up areas, restricted or prohibited areas, and conduct night flights.

12.4 APPLICATIONS

Drones are used in several ways in the entertainment industry, including film and television, music, and live events. Common applications include the following [8]:

- *Film and Television:* Drones are used for aerial photography and videography, allowing filmmakers to capture unique angles and dynamic camera movements, which was previously difficult or expensive to achieve. Drones can also be used for pre- and post-production tasks like scouting remote locations, mapping, and story development. TV commercial directors have been especially quick to adopt the new technology, using drones to film commercials for such brands as Tesla, Chrysler, and Nike. The increasing use of drones is changing the way that movies and TV shows are made.

- *Video Production:* Drones are used in a variety of film and video production jobs, including aerial cinematography. Drones can also capture footage that would otherwise be inaccessible to on-ground camera sets, such as dramatic and cinematic views. Drones are also more affordable than traditional methods, such as using helicopters or cranes, which can be expensive to build, rent, and crew.
- *News:* Many news organizations often use helicopters or planes which have higher costs and require people on-site to operate. Additionally, if a journalist is covering conflict or natural disaster, drones can remove people from that danger while still covering the story.
- *Music Videos:* Music videos have evolved significantly over the years. Drones bring a new dimension to music videos, allowing for creative storytelling and unique perspectives that were previously unattainable. They have been used in concerts to perform backup dancing and other acts. They allow artists to capture aerial shots that soar above cityscapes or natural landscapes. This can create cinematic shots that rival those seen in blockbuster films, and immerse viewers in a virtual reality experience.
- *Aerial Photography:* When you take pictures from any flying object, it is aerial photography. A drone pilot for film and video production all stems from one thing: a passion for aviation, technology, and photography. Traditionally, getting the perfect shot always requires multiple tries and numerous hands on deck. But drone technology has made aerial photography much more accessible and affordable. Drone pilots can capture cinematic views that could never have been imagined in the movie industry some years ago. Perhaps more than any other technological development of the past decade, the aerial photography provided by drone technology is having a massive impact on the way movies are made. Figure 12.3 shows aerial photography using a drones [9].

Figure 12.3 Aerial photography using a drone [9].

- *Live Events:* Major live events such as concerts and sports broadcasting have embraced the use of drones to enhance the visual effects for the audience. Drones can be used to create shows for events like weddings, birthdays, parties, family reunions, graduation ceremonies, and the Olympic Games. Drones can also be used to live-stream events, such as when Sony Pictures used drones to broadcast the Spider-Man World Premiere from above Hollywood Blvd. With advancements in drone technology, live event organizers can now create stunning aerial shots that add a dynamic element to the overall experience for viewers. The integration of drones into live events has revolutionized the way audiences engage with these spectacles, making them more immersive and visually stimulating. Figure 12.4 shows a drone used for a live event [10].

Figure 12.4 A drone used for a live event [10].

- *Entertaining Light Shows:* Drone light shows have captivated audiences worldwide, transforming the night sky into a canvas for spectacular aerial displays. Almost everyone would find the idea of an aerial light show in a theme park wholly unbelievable, but with drones, this is possible. Users can fly drone in the air with various lights attached; aerial light shows are unique and enticing to watch. Companies that use drones for light shows combine art and drone technology to create unique aerial performances for a variety of events, including festivals and corporate gatherings. 500 drones in the sky is kind of the norm these days. An exception to this took place on May 2024, when 5,293 drones flew in the air at once for what marked a Guinness World Record-breaking drone show, depicted in Figure 12.5 [11].

Figure 12.5 Entertaining light show [11].

- *Using Drone to Play Games:* You can use drones for enjoyment. It is common for kids to fly kid's drones indoors. For that purpose, drones can be an excellent avenue to play with your children at home. A flying spinner mini drone flying inside your home can be a spark for having a good time as a family game. These mini drones are a great way to build new memories with your family and can be the best gift for drone enthusiasts. You can purchase flying spinner mini drones and make these as presents, especially to your tech-savvy family members and friends.
- *Sports:* When it comes to live sports coverage, drones offer unique and interesting vantage points that cannot be captured any other way. Drones are becoming a staple in the sports industry, particularly in action sports like surfing, snowboarding, and motocross racing. They can capture real-time footage of athletes in action, providing viewers with a unique perspective. Sports broadcasting has been revolutionized by the integration of drones, offering a dynamic perspective that enhances the viewing experience. Moreover, drones are now a staple in advertising campaigns for sports events. Figure 12.6 shows drone coverage of a football game [12].

Figure 12.6 Drone coverage of a football game [12].

12.5 BENEFITS

Drones have become invaluable tools in the film and entertainment industry, offering numerous advantages over traditional equipment. Due to their ability to shoot from a higher perspective, images taken by drones appear extraordinary. With the cost reduction of drones, and the advances in drone technology, opportunities, accessibility, and capabilities have been greatly enhanced. Drones have also been used in various security plans, particularly in high-profile music festivals and other events with gigantic crowds. Other benefits of drones in entertainment include the following [13,14]:

- *Cost-Effectiveness:* Instead of having to rent out expensive cranes or pay for high-cost helicopter rentals, a production crew can now use a relatively low-cost drone to capture a higher-quality shot on film and save money in the process. The cost for renting a very capable drone might only be 25% of the cost of renting other more expensive equipment for the same purpose. This is a godsend to independent filmmakers with limited resources. Drones are more affordable than traditional aerial filming methods. They can also reduce the need for heavy equipment and manpower.
- *Domestic Production:* Since the FAA approval for use of commercial drones in movie and television productions, many scenes can now be filmed domestically using drones, instead of having to whisk an entire production crew to another country.
- *Time Saving:* The time needed to set up a shot with drones is considerably less than it would be with any other type of equipment.

- *Ubiquitous Utilization:* The drone technology is now seemingly everywhere, both in television and in movie scenes. Because the potential applications for drone videography are virtually limitless, more and more studios are becoming aware of the benefits, and are incorporating them as part of the production process.
- *Flexibility:* Drones can access locations that are difficult or impossible for other equipment, and can fly low to the ground, through tight spaces, or high above landscapes.
- *Virtual Reality:* Drones with 360-degree cameras can create an immersive and interactive VR experience for viewers, transporting audiences to locations and perspectives that were previously unattainable. This technology holds immense potential for documentary film productions and educational content.
- *Safety*: In corporate video production, ensuring the safety of the crew and equipment is paramount. Drones have emerged as a significant advancement in achieving this goal. Drones can reduce the risk to workers by performing tasks that would otherwise require them to enter hazardous areas. Professional drone operators adhere to local regulations and guidelines to ensure the safe and responsible use of drones. By utilizing drones, companies can capture stunning footage without compromising safety. Here are the specific safety benefits of using drones in potentially hazardous filming locations and compare them with traditional filming methods.
- *Accessibility:* One of the most significant advantages of using drones in corporate video production is their remarkable flexibility. Drones can effortlessly navigate various environments, from urban landscapes to rugged terrains, enabling filmmakers to capture footage from virtually any location. Unlike traditional camera setups that are often limited by ground-based constraints, drones can soar above obstacles and provide unique perspectives that enhance the visual appeal of corporate videos.
- *Unique Perspectives:* Incorporating drones into corporate video production offers a revolutionary way to capture stunning aerial shots. Unlike traditional filming methods that rely on cranes or helicopters, drones can effortlessly glide through the air, providing unparalleled access to various altitudes and angles. This flexibility allows for the creation of dynamic and engaging visuals that were previously difficult or impossible to achieve. Drones can capture a variety of angles and perspectives that are difficult or impossible to achieve with traditional cameras.
- *High-Quality Footage:* Drones have revolutionized corporate video production by incorporating advanced camera technologies. These high-tech devices now come equipped with state-of-the-art features that rival traditional filming equipment. Modern drones are fitted with high-resolution cameras, often capable of capturing 4K video, ensuring crystal-clear images and videos.
- *Efficiency:* Drones can cover large areas quickly and safely, which can lead to more efficient use of shooting time. They can streamline the process of capturing aerial footage, reducing setup times and the need for large crews.
- *Environmental Impact:* Drones can reduce the environmental impact of film productions. One of the main environmental concerns associated with drones is their potential to disturb wildlife. The presence of drones can cause stress and anxiety among animals. If you are concerned about the environment and want to contribute to conservation efforts, it is important to consider the potential impact of using drones in remote areas.

12.6 CHALLENGES

While drones offer numerous creative possibilities, filmmakers face challenges and considerations. In spite of the increasingly widespread use of drone technology, critics have raised a number of privacy and security concerns. There is the fear that privately owned drones will pose a security threat to citizens if they venture onto private property. Drones can be vulnerable to poor weather conditions and cannot carry heavy, high-end cameras. Other challenges of drones in entertainment include the following [15]:

- *Regulations:* Filmmakers must navigate complex regulations and obtain permits for drone operations. This includes adhering to no-fly zones, privacy laws, and airspace restrictions. Drone regulation falls under the purview of the US Federal Aviation Administration (FAA), which has detailed instructions regarding the registration and flight of unmanned aerial vehicles. Until recently, the FAA effectively banned the use of drones for commercial purposes because they were viewed as potential aviation safety hazards and threats to national security. The FAA is yet to disclose comprehensive regulations for drone use on private property. However, FAA has authorized more than 200 companies and individuals to operate drones for film production. Under the new FAA rules, drones can be used only on sets that are closed to the public and cannot be operated at night.
- *Safety:* Safety is paramount when flying drones on set. Filmmakers must ensure that drones are operated by trained pilots and that safety protocols are followed. Drones can also help event organizers comply with health and safety protocols.
- *Technical Limitations:* Drones have technical limitations, including flight time, payload capacity, and camera capabilities, especially when it comes to filming high-speed action scenes. Filmmakers must choose the right drone for their specific needs. Limited battery life can be a constraint, requiring careful planning to maximize shooting time.
- *Limited Features:* Most recreational drones have limited features compared to industrial drones. These drones may have limited aerial scope. For example, a recreation drone may detect motion in the air but will not have thermal detection or geotagging ability.
- *More Delicate:* Most drones for recreational purposes can be conducive enough to sustain activities but they are more delicate that industrial drone. It is recommended that you are more careful using them since they do not have extra hardware support.
- *Short Flight Duration:* Battery life affects how long a drone can fly. These should last between twenty and forty minutes, unlike industrial drones that could last for hours in the air. The batteries in recreational drones have a lower power capacity, resulting in shorter flight periods.
- *Security Risks*: There are potential security risks and threats associated with this rapidly advancing technology. Drones, with their ability to fly autonomously and carry out various tasks, have undoubtedly revolutionized industries such as aerial photography, delivery services, and surveillance. However, their capabilities also pose significant security concerns. One of the main security risks is the potential for drones to be used as weapons by individuals or groups with malicious intent. Another security threat is the invasion of privacy. Drones can be vulnerable to hacking, and hackers can use them to acquire private information.
- *Weather:* Drones can be sensitive to wind, rain, and other environmental factors, which can impact their flight and the quality of the footage. They may not be able to maneuver properly or gather reliable data in poor weather conditions.

- *Legal Restrictions*: There are strict rules about where drones can fly, and pilots need to be certified and have permits.
- *Technical Limitations:* Drones have limitations on their payload capacity and the types of cameras and equipment they can carry. Most professional-grade drones have a limited flight time of around 20-30 minutes.
- *Next Generation Challenges:* As industrial applications continue to broaden, reliability and robustness in a wider field of environments become critical factors. Drones need to have their own lighting system, camera system, and sensor base while maintaining stability and a long flight time. Add to these future forward capabilities the addition of artificial intelligence to create a fully autonomous device, and the challenges can quickly multiply. Artificial intelligence will continue to play an increased role in navigation, reducing the need for human intervention.

12.7 CONCLUSION

Over the past decade, drone technology has advanced significantly, making drones affordable and increasingly common. Hollywood has capitalized on the drone craze. Drones are also carving out a space for themselves in the film industry that is entirely separate from traditional Hollywood fare. Only time will tell how drones will continue to change entertainment as we know it. With advancements in drone technology, live event organizers can now create stunning aerial shots that add a dynamic element to the overall experience for viewers.

With constant advancements in technology, drones nowadays are produced to provide boundless camera angles, maximum speed chase scenes, over water and in-between tree scenes, overhead landscape and cityscape views, and more. Photography and cinematography have reached great new heights mainly due to drone technology. Moreover, drones are also cost-effective alternatives for many media companies [16]. Drones in the entertainment industry are here to stay. More information about drones in entertainment can be found in the books in [17,18].

REFERENCES

[1] "Drones in film and entertainment industry," https://www.av8prep.com/aviation-library/part-107-drone/drones-in-film-and-entertainment-industry

[2] M. N. O. Sadiku, P. O. Adebo, and J. O. Sadiku, "Drones in entertainment," *International Journal of Trend in Research and Development*, vol. 11, no. 5, September-October 2024, pp. 79-83.

[3] J. Gang, "Drone use in the entertainment industry and beyond," September 2018. https://thebottomline.as.ucsb.edu/2018/09/drone-use-in-the-entertainment-industry-and-beyond

[4] M. Maher, "The future of drones in photography, film, and video production," https://www.shutterstock.com/blog/the-future-of-drones-photography-film-video-production

[5] "How drones work and how to fly them," May 2024, https://dronelaunchacademy.com/resources/how-do-drones-work/

[6] "What are the main applications of drones?" June 2024, https://www.jouav.com/blog/applications-of-drones.html

[7] "Drones in manufacturing: A game-changer for industry," https://viper-drones.com/industries/infrastructure-drone-use/manufacturing/#:~:text=The%20integration%20of%20drones%20into,on%20manufacturing%20is%20no%20exception.

[8] "UAVs for fun: Recreational drones for entertainment," https://www.zenadrone.com/uavs-for-fun-recreational-drones-for-entertainment/

[9] A. Mudgal, "Learn how to use best drones for aerial photography," https://www.pixpa.com/blog/aerial-photography-using-drones

[10] "3 Ways drones are used to enhance music events' experiences," https://blog.tunedglobal.com/3-ways-drones-are-used-to-enhance-music-events-experiences#:~:text=Drones%20are%20used%20to%20amplify,in%20a%20360%2Ddegree%20format.

[11] S. French, "Light show drones: The most popular drones that entertainment companies are using," https://www.thedronegirl.com/2023/11/02/light-show-drones/#:~:text=UVify%3A%20IFO%20drone,-A%20grid%20layout&text=The%20company%20makes%20a%20range,known%20client%20is%20Sky%20Elements.

[12] K. Dilanian, "NFL, members of Congress alarmed by drone incidents over packed sports stadiums," https://www.nbcnews.com/news/sports/nfl-members-congress-alarmed-drone-incidents-packed-sports-stadiums-rcna119959

[13] "How is drone technology changing the media industry?" February 2024, https://www.linkedin.com/pulse/how-drone-technology-changing-media-industry-appsierra-saqsf#:~:text=With%20the%20advent%20of%20drones,perspectives%20that%20were%20previously%20unattainable.

[14] "6 Benefits of using drones in corporate video production," https://mackmediagroup.com/6-benefits-of-using-drones-in-corporate-video-production/

[15] https://www.roboticstomorrow.com/article/2023/06/challenges-in-drone-technology/20734/

[16] B. Kassaye, "Drones: Creating enhanced aerial panoramas in the media and entertainment industry," July 2022, https://youthtimemag.com/drones-creating-enhanced-aerial-panoramas-in-the-media-and-entertainment-industry/

[17] D. Faust, *Entertainment Drones (Drones: Eyes in the Skies)*. Rosen Publishing Group, 2016.

[18] L. L. Bella, *Drones and Entertainment*. Rosen Young Adult, 2016.

CHAPTER 13

DRONES IN SURVEILLANCE

"Soon it will be possible to assert almost continuous surveillance over every citizen and maintain up-to-date complete files containing even the most personal information about the citizen."

- Zbigniew Brzezinski

13.1 INTRODUCTION

Surveillance is the close observation of a person, group of people, activities, infrastructure, building, etc. for the purpose of managing, influencing, directing, or protecting. There are several different methods of surveillance. Methods include GPS tracking, camera observation, and stake-outs. Traditional observational surveillance methods are typically limited by the stationary nature of the camera, which is usually handled manually or fixed upon a structure. Drones provide the ideal solution to the problems and limitations faced by other surveillance methods. Drone surveillance presents an easier, faster, and cheaper method of data collection. It provides imagery that the human eye is unable to detect.

A drone is an aerial vehicle designed to be used without a human pilot onboard. Drones are being used in a wide variety of applications, from surveilling perimeters to conducting 24/7 surveillance of huge sectors. Drone surveillance is the use of unmanned aerial vehicles (UAVs) to capture still images and video at a high altitude to gather information about specific targets. Surveillance drones have emerged as powerful tools in the realm of aerial monitoring. They have ushered in a new era for private investigators, expanding their capabilities in surveillance investigations beyond traditional means. The integration of drone technology into various sectors has paved the way for innovative solutions to age-old challenges [1].

This paper highlights the applications of drones in surveillance. It begins with describing what a drone is. It covers surveillance drones. It presents some applications of drones in surveillance. It highlights some benefits and challenges of drones in surveillance. The last section concludes with comments.

13.2 WHAT IS A DRONE?

The FAA defines drones, also known as unmanned aerial vehicles (UAVs), as any aircraft system without a flight crew onboard. Drones include flying, floating, and other devices, including unmanned aerial vehicles (UAVs), that can fly independently along set routes using an onboard computer or follow commands transmitted remotely by a pilot on the ground. A typical drone is shown in Figure 13.1 [2]. A drone is usually controlled remotely by a human pilot on the ground, as typically shown in Figure 13.2 [3]. Drones can range in size from large military drones to smaller drones. Drones, previously used for military purposes, have started to be used for civilian purposes since the 2000s. Since then, drones have continued to be used in intelligence, aerial surveillance, search and rescue, reconnaissance, and offensive missions as part of the military Internet of things (IoT). Today, drones are used for different purposes such as aerial photography, surveillance, agriculture, entertainment, healthcare, transportation, law enforcement, etc.

Figure 13.1 A typical drone [2].

Figure 13.2 A drone is usually controlled by operators on the ground [3].

Commercial drones have come a long way in the last decade. Drones work much like other modes of air transportation, such as helicopters and airplanes. When the engine is turned on, it starts up, and the propellers rotate to enable flight. The motors spin the propellers and the propellers push against the air molecules downward, which pulls the drone upwards. Once the drone is flying, it is able to move forward, back, left, and right by spinning each of the propellers at a different speed. Then, the pilot uses the remote control to direct its flight from the ground [4].

Drone laws exist to ensure a high level of safety in the skies, especially near sensitive areas like airports. They also aim to address privacy concerns that arise when camera drones fly in residential areas. These include the requirement to keep your drone within sight at all times when airborne. In the United States, drones weighing less than 250g are exempt from registration with civil aviation authorities. If your drone exceeds 250g in weight, you will also require a Flyer ID, which requires passing a test [6]. It is necessary to register as an operator, be trained as a pilot, and have civil liability insurance, in addition to complying with various flight regulations, and those of the places where their use is permitted.

Most drones have a limited payload, usually under 11 pounds. Drones are classified according to their size. Here are the different drone types:

- Nano Drone: 80-100 mm
- Micro Drone: 100-150 mm
- Small Drone: 150-250 mm
- Medium Drone: 250-400 mm
- Large Drone: 400+ mm

One of the emerging trends in drone use for factories is the utilization of LiDAR technology. LiDAR stands for Light Detection and Ranging. This technology provides accurate depth information essential for understanding the three-dimensional structure of the environment. LiDAR sensors emit laser beams to measure distances to objects, creating high-resolution 3D maps of the surrounding terrain and objects. The ability to capture detailed data through LiDAR technology has opened up opportunities for better predictive maintenance, reduction in inspection times, and overall cost savings [5].

13.3 SURVEILLANCE DRONES

Traditional aerial surveillance with a helicopter achieves the desired result but is also very costly. A surveillance drone is an unmanned aerial vehicle (UAV) equipped with cameras, sensors, and other monitoring devices to gather visual and/or audio information from the environment below. Surveillance drones can be equipped with sophisticated imaging technology that provides the ability to obtain detailed photographs of terrain, people, homes, and even small objects. They can be equipped with different monitoring tools, including high-resolution cameras, thermal sensors, GPS, and sensors to detect motion. They can be remote controlled or fully automated. They keep an eye on operations, individuals, and valued assets from above. Some of these drones are displayed in Figure 13.3 [7]. Figure 13.4 shows how a surveillance drone works [8].

Figure 13.3 Some surveillance drones [7].

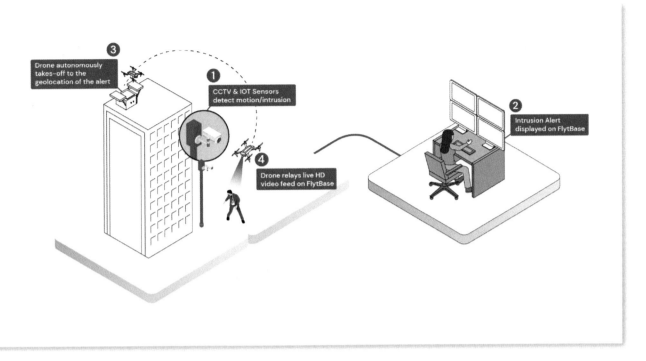

Figure 13.4 How a surveillance drone works [8].

Figure 13.5 A surveillance drone designed for night operations [9].

Unlike some nocturnal animals, such as owls and cats, which see well in the dark, humans need light to distinguish their surroundings, making it almost impossible to discern anything in the dark or at night. Some night surveillance drones have great features or specs to capture nighttime video perfectly. Surveillance drones designed for night operations come with a set of specifications that enable them to function effectively in low-light and challenging conditions. Night vision cameras can be equipped with drones for filming and recording at night. A surveillance drone designed for night operations is typically shown in Figure 13.5 [9]. The primary use of night vision drones is photography. This might include wedding photography, where weddings or receptions and celebrations continue long after dark.

Surveillance drones are of great use in today's life. These drones have rapidly become indispensable tools in a wide range of applications. Their versatility, maneuverability, and ability to capture high-quality imagery from above have opened the door to numerous possibilities across various sectors.

13.4 APPLICATIONS

Recently, surveillance drone technology has been used in different fields, such as military, agriculture, law enforcement, border patrol, etc. Common applications of surveillance drones include the following [10]:

- *Surveying:* Surveillance drones capture high-quality aerial images which provide valuable information about the topography, land features, and vegetation cover. Moreover, drones can cover large areas quickly and efficiently. Thus, they enable rapid data collection compared to ground-based surveys.
- *Agriculture:* Surveillance drones have potential benefits in the agriculture sector. They can assist farmers by providing them with accurate and timely data about soil variability, moisture levels, and nutrient distribution. For example, drones equipped with high-resolution cameras can capture detailed imagery of crop fields. This imagery can be used to monitor crop health, identify areas of stress or disease, and assess overall crop condition.
- *Emergency Response:* In the aftermath of a natural disaster or conflict, drones can be used to provide vital aid and support. Drones can be deployed quickly to assess the situation in emergency scenarios such as natural disasters, accidents, or hazardous incidents. They can be launched within minutes and provide immediate aerial reconnaissance to emergency responders on the ground. Drones can capture high-resolution aerial imagery of disaster-affected areas.
- *Mining:* Drones equipped with specialized sensors, such as LiDAR and multispectral cameras, can conduct aerial surveys of prospective mining sites. Drones play a crucial role in environmental monitoring and compliance for mining operations. They can assess vegetation cover, monitor water quality, and detect changes in land use to ensure compliance with environmental regulations.
- *Aviation:* Drones can be used for perimeter security patrols around airports and aviation facilities. They can monitor fences, gates, and other access points, providing real-time surveillance to detect intruders or suspicious activity and enhance overall security measures. Drones equipped with high-resolution cameras and sensors can conduct inspections of airport runways, taxiways, and other infrastructure.

- *Law Enforcement:* Security drones equipped with live video cameras, infrared cameras, thermal sensors, and LiDAR are used in large quantity by law enforcement agents. Drones can serve to augment human guards by patrolling the worksites and capturing aerial footage of the assets, securing perimeters and preventing break-ins. They can be used for crime scene investigations, monitoring crowds, and border security. Law enforcement must obtain special permission to use drones for surveillance. Privacy law has not kept up with the rapid pace of drone technology, and police may believe they can use drones to spy on citizens with no warrant or legal process whatsoever. There are no systems currently available to law enforcement that can conduct fully autonomous operations. To counter the threat of surveillance, privacy advocates have focused solely on requiring warrants before the use of drones by law enforcement. A police drone is shown in Figure 13.6 [3].

Figure 13.6 A police drone [3].

- *Military:* Military surveillance drones serve a dual purpose. They can provide real-time intelligence on enemy movements and geographic information, which is critical in border surveillance and conflict zones. They also excel in the identification and tracking of potential threats, aiding precision strikes and reducing unintended harm. These capabilities enhance military strategies and safety in the field.
- *Security:* The use of drones in the security industry has been widely debated over the past years. Drones can provide a lot of possibilities as a physical security technology tool. Drones can be used for indoor surveillance in large facilities like warehouses and shopping malls. They can also be used for outdoor security and surveillance missions, such as highway inspection, border patrol, and perimeter security. Drone security patrols have the ability to monitor vast areas and provide high-quality images and video footage in real time, which could also mean that fewer

on-site security officers are needed to protect a site or property. While securing parking lots, drones can compare license plates against those on the hot sheet, in this way helping security staff spot stolen cars or identify unauthorized vehicles.
- *Mapping:* Drones can be used to create accurate maps for strategic purposes, such as mapping dangerous roads. A mapping drone helps visualize and georeference any foreign environment that is expected to be the action ground of defense and security missions. You can scan large regions and create digital surface models and 3D maps for further strategic planning, logistics, and other military intelligence tasks.
- *Maritime Surveillance*: Drones operating in the air and at sea can enhance nations' ability to patrol and control their coastal waters. For example, Africa's coastal nations are responsible for more than 13 million square kilometers of maritime territory. Countries are embracing drones to strengthen their ability to patrol and control their coastal waters because they allow for surveillance and documentation of piracy, illegal fishing, and other activities with minimal human resources.
- *Crowd Monitoring:* In scenarios such as public events or protests, drones can provide aerial views of crowds, assisting security personnel in identifying potential threats or disturbances. Law enforcement agencies can employ drones for surveillance during operations, tracking suspects, or managing emergency situations effectively. Surveillance drones can play a role in monitoring for human rights abuses.

13.5 BENEFITS

Drones make aerial surveillance easy for anyone to do. They allow the real-time monitoring of entire urban populations. Other benefits include [11]:

- *Autonomy:* This is the key benefit that drones integrated with AI can bring to physical security. Human operators involved in drone surveillance are prone to fatigue and errors, can lose concentration, and miss threats. Autonomous drone security systems can be programmed to operate reliably and continuously and can be rapidly deployed across large areas.
- *Monitoring Hazardous Areas:* Drones can do tedious and repetitive tasks of monitoring large areas very efficiently and rapidly, inspecting hard-to-reach locations and gathering data necessary to assess potentially dangerous situations. Due to the ability to cover vast areas regardless of terrain, drones can get closer to hazards, such as high voltage areas, without putting humans at risk of harm and enabling better-informed decisions during adverse incidents.
- *Cost Saving:* While the initial investment in drone technology will be expensive, the long-term benefits far outweigh the initial cost in terms of return on investment. The cost of a security drone is only 20% of the cost of helicopter patrol and 40% of the cost of foot patrol. Drones greatly reduce the cost of aerial surveillance. They are far cheaper to use for aerial surveillance than traditional crewed aircrafts. A drone does not require an expensive certified pilot to operate. Drones can help save cost especially when it comes to large industrial facilities, like oil and gas sites, solar farms, storerooms or securing pipelines where proper video surveillance is critical to reducing risks of damage, leaks and protecting against theft of equipment and raw

materials. However, buying, installing and maintaining cameras in remote areas is expensive. With drones, carrying out inspections is cheaper.

- *Speed:* Drones are much faster than a patrol vehicle or a security officer, which allows them to reach the scene of the incident several times quicker and give the opportunity to provide a rapid remedial response.
- *Efficiency:* The number of drones is constantly increasing every year worldwide, as they have already proven their efficiency when deployed to patrol perimeters, inspect pipelines, deliver medical supplies, and gather data.
- *Rapid Response:* Emergencies require immediate action because every minute can be a difference between life and death. With security drones, the process is much more effective. A drone for surveillance can perform perimeter patrols 30 times faster than a manned patrol.
- *Improved Visual Capacities:* Surveillance drones can fly from a high altitude that allows offering a wide aerial viewpoint without blind spots. With high-quality sensors and HD cameras, drones can detect anomalies or events in low light conditions and from meters away.
- *Safe Missions:* By using remotely controlled drones, you can help reduce the risk to security staff as pilots will be a safe distance away. In the event of a suspect apprehension or disaster outbreak, it is best to send a drone to see the situation first and investigate potential risks before humans get inside.
- *Drone as First Responders* (DFR): The air robots provide an additional set of eyes that can help first responders quickly assess and respond to situations while providing valuable situational awareness. The main goal is to provide immediate assistance and support to those in need, stabilize the situation, provide medical attention and transport, and ensure the safety of everyone involved until more advanced medical or emergency services arrive.
- *Autonomous Deployments:* Autonomous operation of drones can be utilized for 24/7 deployments. An autonomous drone requires no pilot training for security guards and no preexisting flying skills. The drone is deployable in less than 30 seconds and everything from takeoff and landing to path-planning is handled automatically.

13.6 CHALLENGES

Drone use has led to the rise of many security, safety and privacy issues. Drones' characteristics, namely small size, low cost and ease of use, made them a preferred choice for criminals. Another obvious obstacle for driving the broad adoption of drones is hardware (battery life, sensors, cameras and receivers' quality, flight stability, etc.). Drones can malfunction and crash into a nearby house or a group of people, causing property damage and human injuries. The battery is one of the critical components of a drone. The law around conducting aerial surveillance very much remains in a gray area. Other challenges facing surveillance drones include the following [12]:

- *Privacy:* Drones can raise privacy and civil liberties concerns. Surveillance from the skies threatens to end privacy in public. Drones may also not be used for surveillance in violation of another party's reasonable expectation of privacy. People's privacy is at high risk of being exposed by unwanted interference. Drones can record their movement and capture images without their knowledge or permission. Drones and other aerial surveillance technologies

may enable targeted surveillance that protects privacy, while still allowing for the collection of evidence.
- *Regulation:* The integration of drones into the National Airspace requires new regulation to protect people's privacy against aerial surveillance. The Federal Aviation Administration (FAA) has yet to provide information on how these drones will be used. Maintaining compliance with FAA requirements and regulations have been a challenge for security agencies. Previously, such unmanned aircraft were generally not allowed to fly over people, fly at night, or be out of the line of sight of the operator without an FAA waiver. Now, these drones can operate at concerts, sporting events, and for security purposes, although with certain constraints.
- *Legislation:* Drones require new legislation from governments and budgets to pay for equipment and trained personnel. Many critics of drones raise the legitimate concern that the government's collection of aerial imagery and video will enable pervasive surveillance that allows the government to know what all citizens are doing at all points in time. Some jurisdictions have enacted limitations on how information gathered from drones may be used. Legislators should follow a property rights approach to aerial surveillance. The most effective solution is to adopt a property rights approach that does not disrupt the status quo. Legislators should adopt policies that address collection and retention of information in a way that focuses on the information that is collected, how it is stored, and how it is accessed. They should focus on controlling the duration of surveillance. Legislators should craft simple duration based surveillance legislation that addresses the potential for persistent surveillance. Crafting legislation that places aggregate limits on how long law enforcement may surveil specific persons or places can protect against the possibility of persistent surveillance. Surveillance of longer than 48 hours is permissible only when accompanied by a warrant and probable cause. Legislators should adopt transparency and accountability measures. To hold law enforcement accountable, legislators should mandate that the use of all aerial surveillance devices (manned or unmanned) be published on a regular basis (perhaps quarterly) on the website of the agency operating the system.
- *Equipment Breakdown:* Equipment breakdown can lead to property damage, business interruption, and additional expenses.
- *Communication:* Drone communications may be prone to eavesdropping and man-in-the-middle attacks.
- *Limitations:* Drones have a limited flight time. The brick-size batteries used by drones are heavy and get used up quickly. Gasoline engines, meanwhile are noisy and emit exhaust. Drones powered by a hydrogen fuel cell can fly farther and up to three times longer than comparably sized battery-powered aircraft. They operate silently, emit nothing but water, and can be refueled quickly.
- *Harm*: Drones can cause significant harm to people, property, or wider environments, such as when entering the turbines of passenger jet engines or falling from great heights into crowded areas. There are several documented incidents of drones causing bodily harm. Drones are also capable of presenting other forms of unique harm to security. Their use for smuggling illegal contraband into prisons has already been flagged as a serious risk.

13.7 CONCLUSION

Drone surveillance refers to the act of keeping a visual track of an individual, a group, objects, or a situation for the purpose of thwarting any kind of threat. It can survey objects that may be out of reach and can get a first-person view that photographers do not usually get. It provides real-time insight into security and emergency situations for better control, accurate intelligence gathering, comprehensive situational awareness, and more informed decision making. While the idea of deploying a personal drone for surveillance might be enticing, it is imperative to acknowledge the potential legal and ethical pitfalls. Engaging a licensed private investigator offers a multitude of benefits, with their expertise in privacy laws standing out as a key advantage.

While drones may be a great asset to your enterprise, there are still a lot of uncharted waters out there. Drones also are at risk of electronic jamming, hacking, and other countermeasures. Professional drones used for surveillance and security have the capability to disrupt a wide range of industries. The cost of drones is low enough that human rights organizations and even private citizens can buy them. More information about drones in surveillance can be found in the books in [13-20].

REFERENCES

[1] M. N. O. Sadiku, U. C. Chukwu, and J. O. Sadiku, "Drones in surveillance," *Innovative Multdisciplinary Journal of Applied Technology,* vol. 2, no. 10, 2024, pp. 50-60.

[2] "Autonomous security drones,"
https://heighttechnologies.com/autonomous-security-drones/

[3] "The 5 best surveillance drones: Next-level inspection UAVs,"
https://www.zdnet.com/article/best-surveillance-drone/

[4] "How drones work and how to fly them," May 2024,
https://dronelaunchacademy.com/resources/how-do-drones-work/

[5] "What are the main applications of drones?" June 2024,
https://www.jouav.com/blog/applications-of-drones.html

[6] "Drones in manufacturing: A game-changer for industry,"
https://viper-drones.com/industries/infrastructure-drone-use/manufacturing/#:~:text=The%20integration%20of%20drones%20into,on%20manufacturing%20is%20no%20exception.

[7] "Surveillance drones – A complete guide to aerial monitoring," November 2023,
https://www.uasolutions.ch/surveillance-drones/#:~:text=Surveillance%20drones%20have%20various%20capabilities,what%20was%20once%20considered%20possible.

[8] "Drones for surveillance: The ultimate guide,"
https://www.flytbase.com/blog/drone-surveillance-system

[9] "Night vision drone: Most comprehensive guide in 2024," June 2024,
https://www.jouav.com/blog/night-vision-drone.html

[10] "Exploring diverse applications of surveillance drones,"
https://mpowerlithium.com/blogs/blog/exploring-diverse-applications-of-surveillance-drones?srsltid=AfmBOoo_QeoL6PDX782KFGH-2h7ay7mR-TChH_TAyUb8n4Wl1ZVABxSC

[11] "How drones are used to optimize physical security,"

https://www.scylla.ai/how-drones-are-used-to-optimize-physical-security/

[12] "6 challenges that drones bring to perimeter security," https://osltechnology.com/resources/6-challenges-that-drones-bring-to-perimeter-security/

[13] L. Currie-McGhee, *Security and Surveillance Drones.* ReferencePoint Press, 2021.

[14] R. M. Thompson, *Drones in Domestic Surveillance Operations: Fourth Amendment Implications and Legislative Responses.* DIANE Publishing Company, 2012.

[15] V. M. Fair, *Drones and Surveillance.* Mason Crest, 2021.

[16] A. Choi-Fitzpatrick, *The Good Drone: How Social Movements Democratize Surveillance.* MIT Press, 2020.

[17] G. McNeal, *Drones and Aerial Surveillance: Considerations for Legislators.* Brookings, 2014.

[18] S. Kamp, *The Constitutionality of Domestic Surveillance Drones.* Utica College, 2013.

[19] A. Završnik (ed.), *Drones and Unmanned Aerial Systems: Legal And Social Implications for Security and Surveillance.* Springer, 2015.

[20] A. C. Cunningham (ed.), *Drones, Surveillance, and Targeted Killings.* Greenhaven Publishing, 2017.

CHAPTER 14

DRONES IN SPACE EXPLORATION

"A sense of the unknown has always lured mankind and the greatest of the unknowns of today is outer space. The terrors, the joys, and the sense of accomplishment are epitomized in the space program."

– William Shatner

14.1 INTRODUCTION

The vast expanse of outer space has long captured the imagination of humanity. Our solar system, including the sun and everything that orbits it (planets, asteroids, moons, comets, and meteoroids), has attracted more attention from various space agencies compared to other cosmological systems. Various space agencies have invested resources into exploring our solar system physically and through observation. There has been an interest in studying our neighboring planets and moons, such as Venus, Mars, and Titan. Progress in recent technologies has enabled space drones to be considered as valuable platforms for planetary exploration. Due to the advantages of drones compared to other approaches in planetary exploration, ample research has been carried out by different space agencies in the world, including NASA, to apply drones in other solar bodies.

Drones are increasingly being used in space exploration, transportation, and other applications. They are reshaping our understanding of the universe and revolutionizing the space industry by making it more cost-effective, efficient, and accessible. Drones provide an unprecedented capability to traverse difficult terrains and access remote locations, enabling the exploration of regions that were previously inaccessible to surface rovers or landers.

Drones are being utilized across the globe for various diverse uses by commercial, state, military, and individual users. In recent years, there has been a tendency to design and develop concepts of drones and robotic systems for planetary exploration. The integration of drones into space transportation and exploration represents a paradigm shift in how we approach the cosmos. Drones can make space more accessible, cost-effective, and efficient. Scientists use them to complement rovers and other observatory equipment to collect space data. Drones are expected to play a key role in the future of space exploration [1].

This chapter explores major applications of drones in space exploration. It begins with describing what a drone is. It covers space drones. It presents some applications of drones in space exploration. It highlights some benefits and challenges of drones in space exploration. The last section concludes with comments.

14.2 WHAT IS A DRONE?

The FAA defines drones, also known as unmanned aerial vehicles (UAVs), as any aircraft system without a flight crew onboard. Drones include flying, floating, and other devices, including unmanned aerial vehicles (UAVs), that can fly independently along set routes using an onboard computer or follow commands transmitted remotely by a pilot on the ground. A typical drone is shown in Figure 14.1 [2]. A drone is usually controlled remotely by a human pilot on the ground, as typically shown in Figure 14.2 [3]. Drones can range in size from large military drones to smaller drones. Drones, previously used for military purposes, have started to be used for civilian purposes since the 2000s. Since then, drones have continued to be used in intelligence, aerial surveillance, search and rescue, reconnaissance, and offensive missions as part of the military Internet of things (IoT). Today, drones are used for different purposes such as aerial photography, surveillance, agriculture, entertainment, healthcare, transportation, law enforcement, etc.

Figure 14.1 A typical drone [2].

Figure 14.2 A drone is usually controlled by operators on the ground [3].

Commercial drones have come a long way in the last decade. Drones work much like other modes of air transportation, such as helicopters and airplanes. When the engine is turned on, it starts up, and the propellers rotate to enable flight. The motors spin the propellers and the propellers push against the air molecules downward, which pulls the drone upwards. Once the drone is flying, it is able to move forward, back, left, and right by spinning each of the propellers at a different speed. Then, the pilot uses the remote control to direct its flight from the ground [4].

Drone laws exist to ensure a high level of safety in the skies, especially near sensitive areas like airports. They also aim to address privacy concerns that arise when camera drones fly in residential areas. These include the requirement to keep your drone within sight at all times when airborne. In the United States, drones weighing less than 250g are exempt from registration with civil aviation authorities. If your drone exceeds 250g in weight, you will also require a Flyer ID, which requires passing a test [6]. It is necessary to register as an operator, be trained as a pilot, and have civil liability insurance, in addition to complying with various flight regulations, and those of the places where their use is permitted.

Most drones have a limited payload, usually under 11 pounds. Drones are classified according to their size. Here are the different drone types:

- Nano Drone: 80-100 mm
- Micro Drone: 100-150 mm
- Small Drone: 150-250 mm
- Medium Drone: 250-400 mm
- Large Drone: 400+ mm

One of the emerging trends in drone use for factories is the utilization of LiDAR technology. LiDAR stands for Light Detection and Ranging. This technology provides accurate depth information essential for understanding the three-dimensional structure of the environment. LiDAR sensors emit laser beams to measure distances to objects, creating high-resolution 3D maps of the surrounding terrain and objects. The ability to capture detailed data through LiDAR technology has opened up opportunities for better predictive maintenance, reduction in inspection times, and overall cost savings [5].

14.3 SPACE DRONES

Scientists need special tools and devices to explore space. Drones have emerged as useful tools, needed in space to observe and survey areas. Drones help scientists learn more about asteroids as well as other planets and moons. The advantage of using drones is they are cheaper than sending people and have better potential to fly over places the rovers cannot access. A standard drone cannot fly to space independently for many reasons. For the propellers to effectively lift the drone, there must be enough air in the atmosphere. As you move up the Earth's atmosphere, the air becomes thinner and the pressure drops. The reasons drones cannot fly in space include the following [7]:

1. Lack of air push.
2. Lack of movement.
3. Lack of powerful engine.
4. Drones cannot work in vacuum.
5. Propellers cannot work.

That is why NASA is working on a different propulsion system for the drones meant to fly to space. Thus, drones can fly in space but with customization to withstand low atmospheric pressure and weak gravity. A modified or assisted drone can fly to space, although the pilots have to add modified communication equipment to effectively control the drone. To fly in space, the drone needs some assistance in the form of a weather balloon, which allows the drone to get to the incredibly high altitude required in order to be able to see the circumference of planet earth. Once the weather balloon reaches a certain point, the balloon will pop, resulting in the drone falling back to earth.

Space drone is a spacecraft designed to dock with existing satellites that are running low on fuel but are otherwise operational. It uses electric propulsion to maneuver to its target satellite, attaching to the satellite's launch vehicle interface ring via robotic arms.

Solar panels and energy storage systems are helping space drones sustain longer missions. Drones are designed and fabricated with different shapes and sizes and hence various methods and materials for manufacturing. A typical space drone is shown in Figure 14.3 [8].

Figure 14.3 A typical space drone [8].

14.4 APPLICATIONS

In recent years, there has been a paradigm shift in the way we approach space transportation and exploration, with the integration of drones playing a pivotal role. Drone technology has multiple capabilities and applications. Space drones can assist in mapping, sample collection, and geological and atmospheric analysis. Common applications of drones in space exploration include the following [9]:

- *NASA*: NASA is using drone technology for a variety of reasons, such as aerial imagery, inspections, and mapping. Drones are used at NASA's Stennis Space Center to capture images and video and record data. NASA is developing a nuclear rotorcraft drone called Dragonfly that will explore the skies and surface of Titan. At NASA's Stennis Space Center at Mississippi, drones are becoming a go-to resource for use on difficult and potentially dangerous jobs, helping to save time and costs. As NASA integrates drone mapping operations into existing software, future scanning is expected to allow the creation of 2D or 3D real-time maps and models. A team of NASA engineers wants to put drones on Mars. NASA already uses drones in many of their operations. Figure 14.4 shows an astronaut using a digital glove in space to control a drone on the moon [9].

Figure 14.4 A astronaut uses a digital glove in space to control a drone on the moon [10].

- *Aerospace:* Drones are transforming the aerospace industry in many ways. Drones can inspect aircraft, runways, and airport terminals more quickly, precisely, and cost-effectively than traditional methods. For example, Airbus uses drones to photograph the upper part of its A330 aircraft during final inspections. Drones can send service work orders to maintenance teams as soon as they identify a fault. This can reduce maintenance costs and improve safety. The aerospace industry is also exploring the use of space drones to provide inspections of commercial aircraft. With traditional technology, visual inspections can take up to six hours, but space drones could significantly reduce the amount of time and provide increased accuracy and ease of documentation [11]. Figure 14.5 shows a drone flying at the Golden Triangle Regional Airport [12].

Figure 14.5 A drone flying at an airport [12].

- *Satellite Communications*: Satellite communication can provide reliable wireless connectivity for drones almost anywhere in the world. Drones with satellite communication can provide real-time data and imagery to command centers, allowing for better situational awareness and decision-making. They can be used for efficient delivery, inspection, and monitoring services in areas such as e-commerce, logistics, and agriculture. They can also be used for collaborative exploration and experimentation in science, engineering, education, and art. However, satellite communication terminals can be bulky, making them unsuitable for smaller drones. Figure 14.6 shows drones crossing with satellite [13].

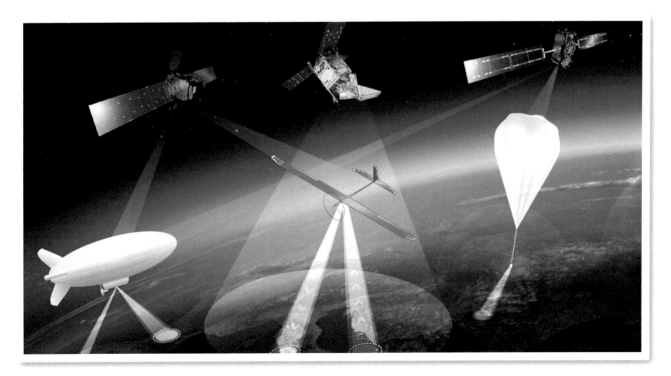

Figure 14.6 Drones crossing with satellites [13].

- *Safety Inspection:* The safety of astronauts and the integrity of space vehicles are paramount. Drones can be used for safety inspections before and during launch. They can check for debris on the launchpad, ensuring a clean and safe departure. Drones equipped with high-resolution cameras and sensors can be used to inspect launch facilities. Additionally, drones can monitor the weather conditions, providing real-time data that helps in making critical launch decisions.
- *Autonomous Spacecraft Repairs:* Space exploration missions are often marred by the challenges of long distances and communication delays. Drones can be deployed to autonomously repair spacecraft, extending their operational lifetimes. In the future, autonomous drones could repair critical systems, such as solar panels or communication antennas, while the spacecraft is far from Earth. Autonomous drones could be equipped with tools and technology to handle routine maintenance and repairs.
- *Aerial Exploration:* On planets like Mars, drones are being used to conduct aerial exploration. For example, the Mars Helicopter has provided stunning imagery and data from the Martian surface, showcasing the potential for drones to reach areas that are inaccessible to traditional rovers.
- *Sample Collection:* Drones can also be used to collect samples from planets and moons. A drone can fly to a specific location, gather samples, and return to the main spacecraft, all while operating autonomously.
- *Resource Survey:* Drones can be deployed to conduct resource surveys on the Moon. They can explore the lunar surface, identify valuable resources like water ice, and even perform preliminary mining activities. This approach could significantly reduce the cost of lunar missions by utilizing local resources.

14.5 BENEFITS

Drones have the potential benefits in terms of cost-efficiency, versatility, and adaptability to different environments. Safety is a significant benefit of drone usage. Drones offer unique capabilities to get close views of potentially life-threatening situations. They can impact almost any business and provide a positive return-on-investment quickly. They are rapidly evolving beyond capturing data into major transportation vehicles and carrying us into their future. Drones can also be autonomous, and some can withstand harsh conditions like extreme temperatures or high levels of radiation. They can play a role in deploying satellites and in maintaining space stations. Other benefits of drones in space exploration include the following [14]:

- *Sample Collection:* Drones can be used to collect samples from planets and moons. A drone can fly to a specific location, gather samples, and return to the main spacecraft, all while operating autonomously. This could be a game-changer for missions aiming to study the composition of extraterrestrial bodies.
- *Planetary Exploration:* Exploring other planets and celestial bodies has always been a fascination for scientists and space agencies. Drones are playing a pivotal role in these endeavors, bringing new dimensions to planetary exploration. They can explore planets like Mars and Venus, and can provide high-resolution information about the planet's surface, atmosphere, and interior.

- *Asteroid Exploration:* Drones can explore and analyze asteroids up close to provide data that helps us understand the early solar system. They can also help identify suitable mining targets.
- *Aerial Exploration:* On planets like Mars, drones are being used to conduct aerial exploration. The Mars Helicopter, also known as Ingenuity, is a prime example. It has provided stunning imagery and data from the Martian surface, showcasing the potential for drones to reach areas that are inaccessible to traditional rovers.
- *Survey:* Drones can be deployed to conduct resource surveys on the Moon. They can explore the lunar surface, identify valuable resources like water ice, and even perform preliminary mining activities. This approach could significantly reduce the cost of lunar missions by utilizing local resources.

- *Infrastructure Development:* Drones can assist in the development of lunar infrastructure. They can be used to transport materials and equipment, construct habitats, and even set up power generation systems. This paves the way for sustainable lunar colonization.

- *Infrastructure Inspection:* Spacecraft and rockets are stored and prepared in massive hangars and launchpads. Drones equipped with high-resolution cameras and sensors can be used to inspect these facilities, ensuring that they are in optimal condition. They can identify issues such as structural damage, leaks, and electrical faults that could jeopardize a mission.

- *Safety Inspections:* The safety of astronauts and the integrity of space vehicles are paramount. Drones can perform safety checks before and during a launch to ensure a safe departure. They can check for debris on the launchpad and monitor weather conditions. They can check for debris on the launchpad, ensuring a clean and safe departure. Additionally, drones can monitor the weather conditions, providing real-time data that helps in making critical launch decisions.

14.6 CHALLENGES

There are challenges and limitations associated with drone technology in space exploration. These include issues related to power supply, materials, communication delays, thermal control, and the need for robust navigation systems in unfamiliar terrains. Organizations like NASA have been trying their best to send drones at the heights, but there are many factors which pulled them down. Drones which are used on Earth cannot be used the same way in space, because we cannot push air over there. It is challenging for an astronaut to control a drone while outside of their ship on the moon's surface. Other challenges include the following [15]:

- *Regulations:* Space is subject to various international treaties and agreements. Space drones must adhere to regulatory frameworks, which can be complex to navigate. Regulations in the industry have helped address safety and security concerns. Drones have been operating under what amount to temporary regulations, created by governing bodies that did not yet have enough data to create fully fleshed-out rules. For example, the FAA treats all UAVs the same under its regulations, from small drones marketed at hobbyists to large, complex UAVs that can transport large parcels over great distances. The main aim of these regulations is to prioritize the safety of the airspace being traveled by traditional manned commercial and private aircraft.

- *Harsh Environments:* Space is a hostile environment with extreme temperatures, radiation, and vacuum conditions. Drones must be designed to withstand these conditions, which can be damaging to both hardware and electronics.
- *Latency:* In planetary exploration, drones often operate on celestial bodies with significant communication delays, making real-time remote piloting challenging. They must operate autonomously to some degree.
- *Complex Navigation:* The navigation and mapping of alien landscapes are complex tasks. Drones need advanced sensors, cameras, and processing capabilities to avoid obstacles and explore efficiently.
- *Limited Payload Capacity:* Space drones must be lightweight, which often limits their payload capacity for scientific instruments or tools.
- *Power Constraints:* Power generation and energy storage are crucial for space drones. Solar panels, nuclear power sources, or other innovative solutions are required for extended missions.
- *Reliability:* The reliability of space drones is paramount, given the costs and risks associated with space missions. Redundancy and fail-safe mechanisms are necessary.
- *Autonomy:* Space missions, especially planetary exploration, can last for months or even years. Drones must be autonomous and resilient over long durations.

Further research and development in this field is needed to overcome these challenges. Figure 14.7 shows the pros and cons of drone technology [16].

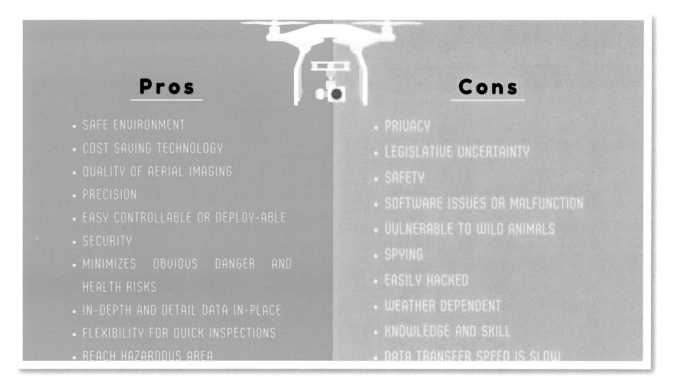

Figure 14.7 The pros and cons of drone technology [16].

14.7 CONCLUSION

As technology continues to advance, we can only expect the role of drones in space exploration to expand further. Drones have the potential to revolutionize future space missions, enabling greater understanding and exploration of the universe. They have proven to be indispensable tools in space exploration and transportation. With their ability to access hard-to-reach areas, conduct autonomous repairs, and gather valuable data, drones are poised to continue transforming our dreams of exploring the cosmos into reality. Although only one drone has been sent to Mars, more research is underway to create better crafts for space exploration.

Drones are a transformative technology that will change in ways we likely never imagined. Much of the future of drones may seem like sci-fi but it i much closer than you may think. Using drones to explore the other planets or moons is one of the main priorities of space agencies. NASA already uses drones in many of their operations, and they hope to use drones widely in the future. More information about drones in space can be found in the books in [17-20] and in the following related journals:

- *Drones*
- *Vertical Space e-Magazine*
- *Progress in Aerospace Sciences*

REFERENCES

[1] M. N. O. Sadiku, U. C. Chukwu, and J. O. Sadiku, "Drones in space," *Innovative Mult-disciplinary Journal of Applied Technology,* vol. 2, no. 10, 2024, pp. 43-49.

[2] T. Bishop, "FAA looking into picture taken by drone above Space Needle," December 2014, https://www.geekwire.com/2014/faa-looking-picture-taken-drone-space-needle/

[3] "The 5 best surveillance drones: Next-level inspection UAVs," https://www.zdnet.com/article/best-surveillance-drone/

[4] "How drones work and how to fly them," May 2024, https://dronelaunchacademy.com/resources/how-do-drones-work/

[5] "What are the main applications of drones?" June 2024, https://www.jouav.com/blog/applications-of-drones.html

[6] "Drones in manufacturing: A game-changer for industry," https://viper-drones.com/industries/infrastructure-drone-use/manufacturing/#:~:text=The%20integration%20of%20drones%20into,on%20manufacturing%20is%20no%20exception.

[7] "Drones in space? Is it possible? Let's find out," March 2020, https://medium.com/srmscro/drones-in-space-is-it-possible-lets-find-out-1031fd726f87

[8] https://www.vectorstock.com/royalty-free-vector/drone-outer-space-earth-vector-8604522

[9] "The use of drones in space transportation and exploration," https://www.av8prep.com/aviation-library/part-107-drone/the-use-of-drones-in-space-transportation-and-exploration#:~:text=Drones%20can%20also%20be%20used,the%20composition%20of%20extraterrestrial%20bodies.

[10] "Astronaut to use a digital glove in space to control a drone on the moon,' November 2019,

[11] https://dronevideos.com/astronaut-to-use-a-digital-glove-in-space-to-control-a-drone-on-the-moon/#:~:text=As%20the%20hand%20and%20fingers,or%20closed%20the%20hand%20is.

[11] "Space drones and the future of drones in aerospace," June 2018, https://www.proponent.com/news/future-of-drones-in-aerospace/#:~:text=The%20aerospace%20industry%20is%20also,accuracy%20and%20ease%20of%20documentation.

[12] B. Guillot and M. Dowell, "Airport benefits of drone technology," April 2019,' https://www.aviationpros.com/aircraft/unmanned/article/12436848/airport-benefits-of-drone-technology

[13] "Crossing drones with satellites: ESA eyes high-altitude aerial platforms," November 2017, https://www.esa.int/Applications/Satellite_navigation/Crossing_drones_with_satellites_ESA_eyes_high-altitude_aerial_platforms#:~:text=Crossing%20drones%20with%20satellites%3A%20ESA%20eyes%20high%2Daltitude%20aerial%20platforms,-28%2F11%2F2017&text=ESA%20is%20considering%20extending%20its,link'%20between%20drones%20and%20satellites.

[14] "The use of drones in space transportation and exploration," https://www.av8prep.com/aviation-library/part-107-drone/the-use-of-drones-in-space-transportation-and-exploration#:~:text=Sample%20Collection,the%20composition%20of%20extraterrestrial%20bodies.

[15] "Challenges and innovations in space-related drone applications," https://www.av8prep.com/aviation-library/part-107-drone/challenges-and-innovations-in-space-related-drone-applications#:~:text=Challenges%20in%20Space%2DRelated%20Drone%20Applications&text=Harsh%20Environments%3A%20Space%20is%20a,to%20both%20hardware%20and%20electronics.

[16] https://equinoxsdrones.com/10-major-pros-cons-of-unmanned-aerial-vehicleuav-drones/

[17] D. R. Faust, *Drones in Space (Drones Are Everywhere!)*. PowerKids Press, 2019.

[18] R. M. Marx, *Creating Space: Drones, Just War, and Jus Ad Vim*. Kent State University, 2016.

[19] M. London, *Space Drones*. Abdo Publishing, 2021.

[20] *Drones in Space: Proceedings of the 32nd Annual Wisconsin Space Conference*, 2023.

CHAPTER 15

DRONES IN THE MILITARY

"To be prepared for war is one of the most effective means of preserving peace."

– George Washington

15.1 INTRODUCTION

The United States has been at the forefront of military drone development and deployment. The number of nations using drones has increased to about 50 in recent years, including China and Iran. The primary purpose of the Department of Defense (DoD) domestic aviation operations are to support Homeland Defense (HD) and Defense Support of Civilian Authorities (DSCA) operations, and military training and exercises. The primary purpose of DoD domestic UAS operations is for DoD forces to gain realistic training experience, test equipment, and tactics in preparation for potential overseas warfighting missions. The vast majority of DoD UAS training is conducted in airspace delegated by the FAA for DoD use [1]. While drones have their many civilian uses in agriculture, education, business, manufacturing, surveillance, film making, etc. military drones are armed ones used in combat.

Drones have rapidly evolved into an essential component of modern warfare. Today, drones primarily serve the military industry around the world. Military drones have become indispensable tools for special operations forces, providing real-time situational awareness, communication relay, and electronic warfare capabilities. They can be used to establish communication networks in the battlefield, relaying signals between ground forces and command centers. Military drone applications have revolutionized the way modern warfare is conducted. From surveillance to combat and logistics, drones have become an indispensable tool for military operations worldwide [2].

This chapter examines the applications of drones in the military. It begins with explaining what a drone is. It discusses military drones. It provides some applications of military drones. It highlights the benefits and challenges of military drones. It concludes with comments.

15.2 WHAT IS A DRONE?

The FAA defines drones, also known as unmanned aerial vehicles (UAVs), as any aircraft system without a flight crew onboard. Drones include flying, floating, and other devices, including unmanned aerial vehicles (UAVs), that can fly independently along set routes using an onboard computer or follow commands transmitted remotely by a pilot on the ground. A typical drone is shown in Figure 15.1 [3]. A drone is usually controlled remotely by a human pilot on the ground, as typically shown in Figure 15.2 [4]. Drones can range in size from large military drones to smaller drones. Drones, previously used for military purposes, have started to be used for civilian purposes since the 2000s. Since then, drones have continued to be used in intelligence, aerial surveillance, search and rescue, reconnaissance, and offensive missions as part of the military Internet of things (IoT). Today, drones are used for different purposes such as aerial photography, surveillance, agriculture, entertainment, healthcare, transportation, law enforcement, etc.

Figure 15.1 A typical drone [3].

Figure 15.2 A drone is usually controlled by operators on the ground [4].

Commercial drones have come a long way in the last decade. Drones work much like other modes of air transportation, such as helicopters and airplanes. When the engine is turned on, it starts up, and the propellers rotate to enable flight. The motors spin the propellers and the propellers push against the air molecules downward, which pulls the drone upwards. Once the drone is flying, it is able to move forward, back, left, and right by spinning each of the propellers at a different speed. Then, the pilot uses the remote control to direct its flight from the ground [5].

Drone laws exist to ensure a high level of safety in the skies, especially near sensitive areas like airports. They also aim to address privacy concerns that arise when camera drones fly in residential areas. These include the requirement to keep your drone within sight at all times when airborne. In the United States, drones weighing less than 250g are exempt from registration with civil aviation authorities. If your drone exceeds 250g in weight, you will also require a Flyer ID, which requires passing a test [6]. It is necessary to register as an operator, be trained as a pilot, and have civil liability insurance, in addition to complying with various flight regulations, and those of the places where their use is permitted.

Most drones have a limited payload, usually under 11 pounds. Drones are classified according to their size. Here are the different drone types:

- Nano Drone: 80-100 mm
- Micro Drone: 100-150 mm
- Small Drone: 150-250 mm
- Medium Drone: 250-400 mm
- Large Drone: 400+ mm

One of the emerging trends in drone use for factories is the utilization of LiDAR technology. LiDAR stands for Light Detection and Ranging. This technology provides accurate depth information essential for understanding the three-dimensional structure of the environment. LiDAR sensors emit laser beams to measure distances to objects, creating high-resolution 3D maps of the surrounding terrain and objects. The ability to capture detailed data through LiDAR technology has opened up opportunities for better predictive maintenance, reduction in inspection times, and overall cost savings [7].

15.3 MILITARY DRONES

The concept of unmanned aerial vehicles(UAVs) dates back to the early 20th century. But it was not until World War II that UAVs began to take shape as a vital military tool. In the late 20th century and early 21st century, military drones evolved rapidly. Today, military drones are used by numerous nations for intelligence, surveillance, and reconnaissance (ISR) missions. In 1959, as tensions between the US and the Soviet Union began to skyrocket, so did drone innovation. The first military conflict in which UAVs played a major role was the Gulf War, after which military drones became commonplace worldwide.

Figure 15.3 A typical military drone [8].

Military drones are remote-controlled aircraft of various sizes designed to perform tasks deemed too dull, dirty, or dangerous for human troops. The main selling point of a military drone is the lack of an on board human pilot. The drones have revolutionized the way modern warfare is conducted by providing a means of engaging targets with precision and reduced risk to military personnel. They are often used in dangerous areas where human soldiers might be at risk. These drones can conduct precision strikes on high-value targets. A typical military drone is shown in Figure 15.3 [8]. Military drones are remotely-piloted UAVs used for monitoring, mapping, target acquisition, intelligence, battle damage management, and surveillance. They are armed with missiles, bombs, or anti-tank weapons. These drones have been a valuable asset in the military for many years. Pentagon takes the issue of military drones used on American soil very seriously.

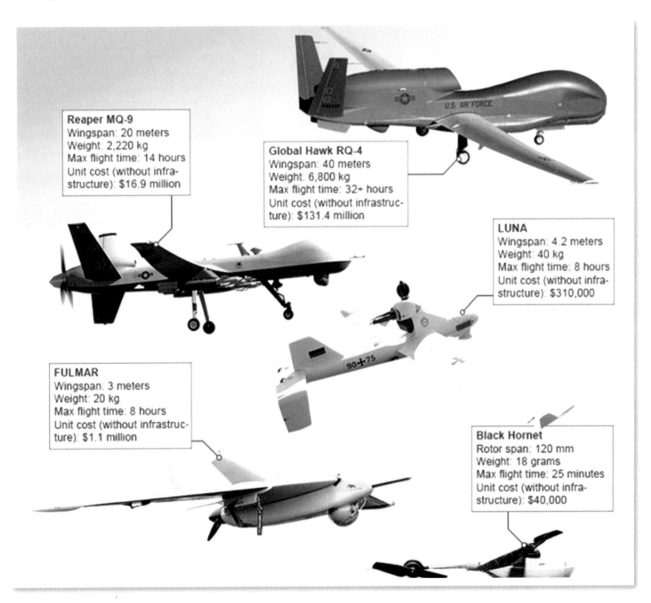

Figure 15.4 Different types of military drones [9].

Figure 15.4 shows different types of military drones [9]. The types include the following [10]:

- Combat drones (or fighter drones) are designed for offensive operations.

- Tactical drones are used in a range of specialized missions, such as communication relay, electronic warfare, and counter-drone operations. Figure 15.5 shows an example of tactical drone [11].
- Transport drones are designed for logistical support, such as delivering supplies, medical evacuation, and personnel transport.
- Reconnaissance drones can be used to monitor enemy movements, troop deployments, and infrastructure.
- Surveillance drones have the ability to observe enemy activities without putting human lives at risk.
- Detecting and tracking drones can provide real-time information on drone locations, flight paths, and potential payloads.
- Swarming drones refer to coordinated deployment of multiple drones to carry out complex tasks or missions.
- Logistics drones are drones that are used for logistical purposes, such as transporting supplies and equipment to troops in the field.
- Target drones are drones that are used as targets for training exercises.
- Stealth drones are drones that are designed to be stealthy and difficult to detect.

Figure 15.5 An example of tactical drone [11].

Leading manufacturers in the global market of military drones include [12]:

1. General Atomics Aeronautical Systems, Inc.
2. Northrop Grumman Corporation
3. Israel Aerospace Industries Ltd
4. BAE Systems Plc
5. Lockheed Martin Corporation

6. Raytheon Company
7. Insitu Inc.
8. AeroVironment, Inc.
9. Turkish Aerospace Industries, Inc.
10. Elbit System Ltd
11. The Boeing Company
12. Thales Group
13. Saab Group

Figure 15.6 shows 3D-printed military drones [13].

Figure 15.6 3D-printed military drones [13].

15.4 APPLICATIONS

Military drone applications have expanded dramatically, playing a critical role in intelligence gathering, combat operations, and logistical support. Drones are used by the military for a variety of purposes, including surveillance, target acquisition, reconnaissance, logistics, and targeted strikes. Common applications of drones in the military include the following:

- *Electronic Warfare:* Electronic warfare is the art of locating enemy forces by the signals that they send out and then isolating them by jamming their communications. This is a critical aspect of modern military operations, and tactical drones have emerged as valuable assets in this domain. They can be equipped with electronic warfare systems to detect, identify, and disrupt enemy radar and communication systems, rendering them ineffective. Drones can jam enemy communications and detect radar installations.

- *Enforcement and Countermeasures:* Law enforcement agencies are also investing in counter-drone technologies and capabilities to detect and neutralize unauthorized or malicious drone activities. These efforts include jamming and spoofing. Once a drone has been detected, jamming and spoofing techniques can be used to disrupt its communication and control systems. Jamming involves the transmission of radio frequency signals to interfere with the drone's remote control or GPS navigation, while spoofing involves sending false signals to deceive the drone's GPS receiver, causing it to deviate from its intended course or crash.
- *Surveillance:* Drones' surveillance capabilities allow for better situational awareness on the battlefield, helping to detect potential threats and plan effective military strategies. In urban warfare scenarios, where the environment is complex and unpredictable, drones can be invaluable assets, assisting ground troops in navigating and understanding the terrain from above.
- *Reconnaissance:* Drones can conduct surveillance missions by hovering over an area for an extended period. They can survey hostile territories and provide real-time information to command centers. This can help protect soldiers from danger and improve situational awareness. They can provide rapid assistance that significantly impacts the survival and recovery of injured individuals. They can help swiftly locate and assess injured soldiers in combat zones.
- *Logistics*: Drones are revolutionizing the delivery of medical supplies. They can carry essential medical equipment and medications to injured personnel in areas inaccessible by traditional means. They can deliver supplies to front-line units in hard-to-reach areas. They can also help evacuate injured personnel. A drone can hover around and land anywhere without using runways. Examples are retail companies, parcel services, and pharmacies using drone delivery to bring goods and medications to our doorstep. Manufacturers are also using drones to ship supplies directly to businesses. Figure 15.7 shows a delivery drone [14].

Figure 15.7 A delivery drone [14].

- *Communication:* Drones can act as airborne communication relays to connect units and command centers, especially in areas with compromised communication infrastructure. Drones also use radio to send data like videos and receive remote control.
- *Search and Rescue:* Drones can be used for combat search and rescue. For example, imagine a search and rescue mission in a dense forest, where a drone uses its AI edge computing capabilities to navigate through the complex environment. Some drones for rescue missions are portrayed in Figure 15.8 [15].

Figure 15.8 Drones for rescue missions [15].

- *Autonomous Navigation:* Modern drones are equipped with cutting-edge AI and edge computing that enables them to navigate complex environments autonomously. This includes obstacle avoidance, terrain analysis, and even adaptive mission planning based on real-time data. It can identify and avoid obstacles like trees and cliffs, analyze the terrain to find the safest and most efficient routes, and adapt its flight plan in real-time based on changes in the environment.
- *Decision Making:* AI algorithms allow drones to make critical decisions rapidly. For instance, a drone can analyze threats, select targets, and even choose flight paths with minimal human input. In a military operation, a drone equipped with AI algorithms can rapidly analyze aerial footage to identify potential threats, such as enemy combatants or unsecured territories. It can then autonomously select the safest flight path to avoid detection or confrontation.
- *Coordinated Operations:* Swarm technology involves multiple drones working together, coordinated through advanced AI. This approach allows for complex, large-scale operations where drones collaborate to achieve common objectives. Swarm drones can be used in various scenarios, from reconnaissance missions to creating real-time 3D maps of battlefields to overwhelming enemy defenses. Figure 15.9 displays a swarm of drones [3].

- *Military Exercises:* Drones are becoming increasingly important in modern military exercises. Military exercises are an important part of military training around the world. It is a way for soldiers to hone their skills, test their equipment and improve their tactics. In recent years, drone technology has played a key role in military exercises. Drones are a game-changer when it comes to military operations and can help soldiers in numerous ways. Drones are also being used in more specialized training exercises. They can be used to train in complex and dangerous environments that would be too dangerous or expensive to train in with live troops. The use of drones in military exercises is not without its challenges. One challenge is that drones can be easily hacked or jammed.

Figure 15.9 A swarm of drones [3].

15.5 BENEFITS

Drones play a crucial role in war, medicine, and rescue missions. Their success in different operations makes them an invaluable asset. Drone strikes ensure the safety of the United States by deconstructing terrorist organizations anywhere in the world. They protect more US military personnel and kill fewer civilians than any other weapon used by the military. They are cheaper than manned aerial warfare or ground combat. The Pentagon has deployed drones to spy over US territory for non-military missions over the past decade. Other benefits of military drones include the following [16]:

- *Better Reconnaissance:* Drones provide real-time information on targets' positions, terrain, and enemy movements to commanders on the ground. Compared to high-altitude aircraft, drones can take closer footage without compromising the quality of both photos and video.
- *Reduced Cost:* Drones are cheaper than conventional aircraft in terms of both price and maintenance. Because drones are unmanned, they also reduce the risk of pilots being injured mid-flight. States and non-state groups that cannot afford to buy fighter jets can buy drones.
- *Increased Convenience:* Compared to conventional aircraft, drones are faster and easier to deploy. They are easier to operate and do not need training as extensive as most aircraft. Also, many drones do not need a runway, and other types can easily fit in a backpack.
- *Enhanced Safety:* Military drones have changed how military bases secure their perimeters and monitor for threats, improving surveillance and soldier safety. Drone operators can provide real-time information without putting themselves at risk. On top of this, that same information also informs commanders where to position their troops to ensure safety. Everything the FAA does is focused on ensuring the safety of the nation's aviation system.
- *Increased Flexibility:* Military forces always need to be ready for anything at a moment's notice. Drones are helpful in readiness. They can even be fully automated. They provide many benefits and advantages that make them extremely useful for different roles.
- *Precision Warfare:* Armed drones have become a game-changer in modern warfare. With the ability to carry various types of munitions, including missiles and guided bombs, they offer precise and lethal firepower. Precision warfare, made possible by drone technology, minimizes collateral damage and civilian casualties, making it a more ethically acceptable form of warfare compared to traditional indiscriminate bombardments.

15.6 CHALLENGES

Critics of the drone strikes argue that drone strikes wreak havoc on civilian communities and in turn create more terrorists than they set out to destroy. Drone operations are secretive, lack sufficient legal oversight, and prevent citizens from holding their leaders accountable. In spite of the growing use of drones, they remain a controversial and unpopular tactic. Other challenges of military drones include the following [16]:

- *Cost:* The Pentagon abhors cheapness; no production line exists for cheap drones or cheap artillery shells. The defense industry prices are prohibitively high. Official procurement figures are classified, but press reports indicate per unit costs for military drones vary from $6,000 to $58,000—— twelve to one hundred times more expensive than Ukraine's home-assembled drones. (Ukraine is producing one million drones). The same cost disparity affects defense just as much as offense on land. Another challenge is that drones can be expensive to operate.
- *Drone Strikes:* Drone strikes violate the sovereignty of other countries and are extremely unpopular in the affected countries. They often raise complex legal and moral dilemmas. They are illegal under both international and United States humanitarian law, which states that lethal force is only permissible when the target poses an immediate threat to the country's survival. And it can be argued that not all drone targets fit this category. The death toll from American

drone strikes was approximately 2,400 in total from 2009-2014 and has risen to more than 6,000 since 2015.
- *War:* Whichever side you are fighting on, war is horrific and lamentable. Less of them are in harm's way because of the use of drones. The proliferation of drone technology poses security risks, as non-state actors and adversarial nations can utilize drones for malicious purposes. Drones have been linked with civilian deaths in many conflict zones
- *Ethical Concern:* The use of military drones, particularly in armed strikes, has raised several legal and ethical concerns. The concerns refer to the use of combat drones in targeted killings, that is the intentional killing of specific individuals outside of an active battlefield, resulting in the unintentional deaths of innocent civilians.
- *Accountability:* Another significant concern is the lack of transparency and accountability surrounding drone operations. Due to the covert nature of drone warfare, it can be challenging to ascertain the precise circumstances and justifications for drone strikes. This lack of accountability has led to calls for greater oversight of drone operations to ensure that they are conducted in accordance with international law and ethical standards.
- *Regulations:* To prevent the misuse of drones and ensure public safety, governments worldwide have implemented legal restrictions and regulations governing the use of civilian drones. These regulations typically include requirements for registration, flight restrictions in specific areas (e.g. near airports, military installations, or populated areas), and limitations on drone size, weight, and capabilities.

15.7 CONCLUSION

The rise of drones has been nothing short of revolutionary in the military industry. Modern warfare and national defense using drones make for big business the world over.

With the largest military budget in the world, the United States is a leader in the development and production of the drones.

As technology continues to evolve, so too will military drone applications. As the use of military drones becomes more prevalent, the development of counter-drone measures and defense systems has become increasingly important. An increase in terror threats, unconventional military threats, and geopolitical tensions worldwide have led to an increase in demand for unmanned aerial vehicles (UAVs) to target terrorist and insurgent groups across the globe. As the market for military drones continues to expand, driven by increasing government funding and technological innovations, drones are set to play an even more crucial role in shaping the landscape of global security and defense strategies.

The search for small disposable and accessible drones in military and defense is on the rise around the world. Every military specialist agrees that Unmanned Aerial Vehicles, or drones, are the future of warfare. Future drones will be even more autonomous, capable of performing complex tasks such as tactical strikes, surveillance missions, and supply deliveries without human intervention. More information about drones in the military can be found in the books in [17-34] and a related journal: *Drones*.

REFERENCES

[1] "Unmanned aircraft systems (UAS)," https://dod.defense.gov/UAS/

[2] M. N. O. Sadiku, P. A. Adekunte, and J. O. Sadiku, "Drones in the military," *International Journal of Trend in Scientific Research and Development*, vol. 8, no. 5, September-October 2024, pp. 58-65.

[3] "The future of warfare and security: Drones in military and HLS applications," January 2024, https://www.maris-tech.com/blog/the-future-of-warfare-and-security-drones-in-military-and-hls-applications/

[4] "The impact of drones on future of military warfare," https://media.inti.asia/read/the-impact-of-drones-on-future-of-military-warfare#:~:text=Here%20are%20some%20additional%20thoughts,each%20other%20in%20real%20time.

[5] "How drones work and how to fly them," May 2024, https://dronelaunchacademy.com/resources/how-do-drones-work/

[6] "What are the main applications of drones?" June 2024, https://www.jouav.com/blog/applications-of-drones.html

[7] "Drones in manufacturing: A game-changer for industry," https://viper-drones.com/industries/infrastructure-drone-use/manufacturing/#:~:text=The%20integration%20of%20drones%20into,on%20manufacturing%20is%20no%20exception.

[8] "Military drones UAS & UAVS for ISTAR," https://heighttechnologies.com/military-istar-drones/

[9] B. Knight, "A guide to military drones," June 2017, https://www.dw.com/en/a-guide-to-military-drones/a-39441185

[10] C. Guarnera, "Overview of military drone applications," April 2023, https://www.bluefalconaerial.com/overview-of-military-drone-applications/

[11] "Decoding military surveillance drones: the ultimate guide," https://elistair.com/military-surveillance-drones/

[12] "Top 13 military drone manufacturers in the world," January 2022, https://roboticsbiz.com/top-13-military-drone-manufacturers-in-the-world/

[13] "3D-printed military drones 'assembled in a matter of hours'," February 2023, https://www.imeche.org/news/news-article/3d-printed-military-drones-assembled-in-a-matter-of-hours

[14] "Advanced air mobility," https://www.gore.com/products/industries/aerospace/advanced-air-mobility?xcmp=aam_aero_ppc_defense_google_na%7c%7cgeneral&s_kwcid=AL!13180!3!647674247881!p!!g!!advanced%20air%20mobility&gad_source=1&gclid=EAIaIQobChMIh8O_x_efiAMVdm0PAh30cSprEAAYAiAAEgLTTfD_BwE

[15] A. Maltsev, "Drones at war and computer vision," December 2023, https://medium.com/@zlodeibaal/drones-at-war-and-computer-vision-a16b8063be7b

[16] M. Abunuwara, "Military drones," https://militarymortgagecenter.com/us-military/aircraft/military-drones/

[17] H. Marcovitz, *Military Drones (World of Drones)*. Referencepoint Press, 2020.

[18] P. J. Springer, *Military Robots and Drones: A Reference Handbook*. Bloomsbury Publishing, 2013.

[19] T. Cooke, *A Timeline of Military Robots and Drones.* Capstone, 2017.

[20] M. Scheff, *Military Drones.* Raintree, 2019.

[21] J. E. Jackson, *One Nation Under Drones: Legality, Morality, and Utility of Unmanned Combat Systems.* Naval Institute Press, 2018.

[22] C. Enemark, *Armed Drones and the Ethics of War: Military Virtue in a Post-heroic Age.* Taylor & Francis, 2013.

[23] A. Završnik, *Drones and Unmanned Aerial Systems: Legal and Social Implications for Security and Surveillance.* Springer, 2015.

[24] D. Cortright, K. Wall, and R. Fairhurst (eds.), *Drones and the Future of Armed Conflict: Ethical, Legal, and Strategic Implications.* University of Chicago Press, 2017.

[25] J. Kaag and S. Kreps, *Drone Warfare.* Polity Press, 20115.

[26] A. Stilwell, *Military Drones: Unmanned Aerial Vehicles (UAV).* Amber Books Limited, 2023.

[27] S. J. Frantzman, *The Drone Wars: Pioneers, Killing Machines, Artificial Intelligence, and the Battle for the Future.* Bombardier Books, 2021.

[28] D. Sloggett, *Drone Warfare: The Development of Unmanned Aerial Conflict.* Skyhorse Publishing, 2015.

[29] E. D. Nucci and F. S. de Sio, *Drones and Responsibility: Legal, Philosophical and Socio-Technical Perspectives on Remotely Controlled Weapons.* Taylor & Francis, 2016.

[30] M. Schuh, *Military Drones and Robots.* Capstone, 2022.

[31] D. R. Faust, *Military Drones.* Rosen Publishing Group, 2015.

[32] C. P. McCarthy, *Military Drones.* ABDO Publishing Company, 2020.

[33] J. I. Walsh and M. Schulzke, *Drones and Support for the Use of Force.* University of Michigan Press, 2018.

[34] M. Chandler, *Military Drones.* Capstone, 2019.

INDEX

Symbols

3D printing drones, 71

A

Accessibility, 11,94,105,143
Accountability, 182
Accuracy, 72,105
Aerospace, 164
Africa, 49
Agriculture, 6,14,29,152
Asset management, 33
Automation, 71,105
Autonomy, 154,168
Aviation, 152
Awareness, 63

B

Bad reputation, 35
Business, 8

C

Care delivery, 47
China, 36,50
Climate, 96
Climate change, 22
Coding, 58,62
Collaboration, 62,72,82
Collision avoidance, 74
Communication, 119,156,179
Competitive advantage, 117
Complexity, 132
Compliance, 72
Connectivity, 24,115
Construction, 31
Corona discharge, 104
Cost, 73,95,181
Cost savings, 105,115,154
Cost-effectiveness, 11,93,107,129,142

Countermeasures, 178
Crime analysis, 127
Critical thinking, 62
Crowd monitoring, 127
Cybersecurity, 73

D

Damage assessment, 128
Data collection, 82,91,93,105,116
Data management, 95
Data retention, 132
Decision-making, 105,109
Delivery, 71,93
Department of Defense (DoD), 171Drone attacks, 106
Drone delivery, 33
Drone strikes, 181
Drone surveying, 34
Drones, 1-12
    Applications of, 6,20
    Benefits of, 11
    Challenges of, 11
    Concept of, 44
    Definition of, 2,15,28,56,67,77,99,110,122,135,148,160,171
    Types of, 3,17
Drones in agriculture, 14-25
    Applications of, 20,21
    Benefits of, 22
    Challenges of, 24
Drones in business, 27-40
    Applications, 29
    Benefits of, 33
    Challenges, 34
    Global uses of, 35
Drones in construction, 76-84
    Applications of, 79
    Benefits, 81
    Challenges, 82
Drones in education, 55-64

Applications of, 57,59
Benefits of, 61
Challenges of, 63
Drones in entertainment, 135-145
    Applications of, 138
    Benefits of, 142
    Challenges of, 144
Drones in law enforcement, 121-133
    Applications of, 126
    Benefits of, 129
    Challenges of, 131
Drones in manufacturing, 65-74
    Applications of, 69
    Benefits of, 71
    Challenges of, 72
Drones in oil & gas industry, 85-96
    Applications of, 89
    Benefits of, 92
    Challenges of, 94
Drones in power systems, 97-107
    Application of, 101
    Benefits of, 104
    Challenges of, 106
Drones in space exploration, 159-168
    Applications of, 163
    Benefits of, 166
    Challenges of, 167
Drones in surveillance, 147-157
    Applications of, 152
    Benefits of, 154
    Challenges of, 155
Drones in telecommunications, 109-119
    Applications of, 113
    Benefits of, 115
    Challenges of, 118
Drones in the military, 171-182
    Applications of, 177
    Benefits of, 180
    Challenges of, 181

E

Efficiency, 11,22,72,82,92,105,116,130,143,155
Electronic warfare, 177
Emergency, 48
Emergency response, 93,106,115,152
Entertainment, 10
Entertainment drones, 138
Environmental concerns, 96
Environmental benefits, 117
Environmental impact, 22,63,94,143
Environmental monitoring, 91
Environmental regulation, 93

Environmental science, 58
Equipment tracking, 80
Ethical concern, 182
Europe, 36
Evidence collection, 130
Exploration, 166,167

F

Federal Aviation Administration (FAA), 2,52,77,86,122
Film, 138
Flexibility, 143,181

G

Games, 141

H

Hacking, 12,52,83,119
Harm, 156
Harsh environment, 168
Hatred for drones, 118
Healthcare, 6,43
Healthcare drones, 43-53
    Applications of, 46
    Benefits of, 50
    Challenges of, 51
    Global type of, 49
High costs, 24
Higher education, 60

I

India, 38,50
Industry giants, 93
Inspection, 90,92,102,105,113,117
Insurance, 31
Interoperability, 73
Inventory management, 69
Israel, 38

K

K-23 schools, 59

L

Land surveys, 79
Latency, 168
Law enforcement, 7,152
Leak detection, 91
Legislation, 156
Liability, 83
Light shows, 140
Limitations, 156
Live events, 140

Logistics, 10,119,178

M

Maintenance, 33,69,103,117
Manufacturing, 9
Mapping, 114,154
Mathematical skills, 62
Media, 57
Military, 9,153
Military drones, 174
Military exercises, 180
Mining, 152
Monitoring, 92,94,130,154

N

NASA, 159,162,163
Navigation, 179
Night vision, 130
Noise,12

O

Oil and gas industry, 85,88
Outer space, 159
Overhead lines, 102

P

Package delivery, 32
Pakistan, 50
Photogrammetry, 58
Photography, 8,31,139
Physics, 62
Pilot's responsibilities, 132
Police drones, 124
Power, 35
Power sector, 98
Power Systems, 8
Privacy, 35,51,73,83,95,118,155
Privacy rights, 131
Public perception, 132

Q

Quality control, 69,73

R

Real estate, 31
Reconnaissance, 178,181
Reduced labor, 82
Regulations, 12,24,35,63,73,83,94,106,118,131,144,156,167,182
Reliability, 168
Renewable energy, 104

Rescue operations, 114
Resistance to change, 107
Restaurants, 30
Restrictions, 132
Risks, 24,117,130,144
Road construction, 80
Roof inspections, 81
Rural healthcare, 47

S

Safety, 11,33,35,51,63,72,73,80,82,83,92,95,105,106,115,143,144,166,180
Satellite communications, 165
Search and rescue operations, 127
Security, 11,12,23,32,35,81,91,106,144,153
Signal interference, 119
Singapore, 50
Skill shortage, 119
Smart grids, 104
Social learning, 59
Soft skills, 58
Space, 9
Space drones, 162
Speed, 81,155
Sports, 141
STEM, 55,56,62
Supply chain management, 73
Surveillance, 69,130,154,147,178
Surveillance drones, 150
Surveying, 114,152,167

T

Teacher skills, 63
Technical limitations, 52,95,144
Telecomm drones, 112
Telecommunications industry, 109
Television, 138
Tethered drones, 128
Time saving, 105,142
Traffic management, 127
Training, 107
Transportation, 72

U

United Kingdom, 36
United States, 36,50,88,171
Unmanned aerial vehicles (UAV), 1,27,56,66,67,77,86,121,122,174
Utility companies, 32

V

Virtual reality, 143

W

War, 182
Warfare, 181
Weaponization, 132
Weather, 83, 144

Printed in the United States
by Baker & Taylor Publisher Services